The Politics of Fracking

Over the last decade, the oil and gas industry has garnered a lot of support from the United States federal and state governments in the name of energy independence and economic prosperity. More specifically, hydraulic fracturing or fracking is said to not only make the production of affordable energy possible but also reduce emissions of carbon dioxide by substituting coal with natural gas in the utility sector. Behind the façade of many socioeconomic and political benefits, the process of fracking causes serious environmental concerns. Dismissing the negative externalities of fracking simply raises the question, to what extent have communities close to fracking sites been adversely impacted by it?

In this book, Sarmistha R. Majumdar studies four communities close to fracking well sites in Texas to help illustrate to what extent fracking regulations have been developed in Texas and how effective these regulations have been in safeguarding the interests of individuals in local communities amidst the lure of economic gains from the extraction of oil and natural gas from shale formations. Majumdar has developed a model to show stage by stage community actions to regain their quality of life and the consequences of their actions, if any, on state and local regulations and ordinances, and the oil and gas industry.

This book will be an important resource for scholars of environmental and natural resource politics and policy in the United States.

Sarmistha R. Majumdar is an associate professor at Texas Southern University where she specializes in the analysis of public policies, mainly those related to transportation and environment. She teaches classes on public policies, research methods, and public administration.

Routledge Research in Public Administration and Public Policy

The Politics of Fracking

Regulatory Policy and Local Community Responses to Environmental Concerns

Sarmistha R. Majumdar

Routledge
Taylor & Francis Group

LONDON AND NEW YORK

First published 2019 by Routledge

2 Park Square, Milton Park, Abingdon, Oxfordshire OX14 4RN
52 Vanderbilt Avenue, New York, NY 10017

Routledge is an imprint of the Taylor & Francis Group, an informa business

Library of Congress Cataloging-in-Publication Data
Names: Majumdar, Sarmistha R., author.
Title: The politics of fracking : regulatory policy and local community responses to environmental concerns / Sarmistha R. Majumdar.
Description: New York, NY : Routledge, 2019. | Series: Routledge research in public administration and public policy | Includes bibliographical references and index.
Identifiers: LCCN 2018018579| ISBN 9781138682597 (hardback) | ISBN 9781134823437 (webPDF) | ISBN 9781134823505 (ePub) | ISBN 9781134823574 (MobiPocket/Kindle)
Subjects: LCSH: Hydraulic fracturing–Political aspects–United States. | Hydraulic fracturing–Public opinion–United States. | Energy policy–United States. | Gas industry–United States.
Classification: LCC TN871.255 .M35 2019 | DDC 338.2/7280973–dc23
LC record available at https://lccn.loc.gov/2018018579

ISBN: 978-1-138-68259-7 (hbk)
ISBN: 978-0-367-66540-1 (pbk)

Typeset in Sabon
by Wearset Ltd, Boldon, Tyne and Wear

Contents

Figures

Tables

Preface

Living in the energy capital of the world, where the headquarters and offices of many major oil and natural gas companies are located, it is difficult to remain oblivious to news on the latest trend in oil and gas exploration and production in the state of Texas and the nation. Here, the local newspaper, the *Houston Chronicle* has a special section called "Fuel Fix." In this section, one can find not only the latest rig counts in the state and the country but also news on various aspects of fracking and how it has impacted on local economies and people's lives. As a daily online reader of this local newspaper, it was here that I first got acquainted with the word and process of fracking.

As I followed the news articles on fracking from the east to west coast and in Texas and did my research on the topic, I became all the more convinced that more research and publications on the topic were needed in the field of social science. True, there exist many books and research articles on the geological and engineering aspects of this advanced drilling procedure but only a few such peer reviewed articles and books are available analyzing fracking through the lens of social science. As I was getting ready for an in-depth analysis on the social aspects of fracking, I availed an offer to write a book on it. With my interest and background in public policies and the environment, I decided to focus on the existing regulations on fracking and its foreseen and unforeseen impacts on those communities that are located at close proximity to fracking sites.

As I started writing the book, I made a few visits to fracking sites in the Barnett Shale of North Texas and came into contact with many people who willingly shared with me their stories and expressed their concerns and opinions on this advanced drilling procedure. To all these people, I convey my sincere thanks for providing me with valuable information. I also would like to acknowledge the help that I received from various public agencies throughout the state and the nation in providing me with the facts and statistics. I appreciate the assistance that I received from the journalists of the *Houston Chronicle* and other local newspapers of the state who provided me with much guidance in my quest for more information on the topic. My sincere thanks goes out to my book editor Natalia Mortensen of

Routledge/Taylor and Francis for encouraging me to take up this journey and to Dr David Davis for providing me with insightful comments on my writings. Also, I convey my thanks to Dr Robert Bullard and Dr Michael Adams of Texas Southern University for granting me some relief time from my regular academic duties and checking on my progress, to Texas Southern University's Grants Office for providing me with a seed grant to complete my research project, to my student assistant Youssouf Toure for his comments and criticisms. Finally, I'm grateful to my family for their patience and understanding when I spent long hours on research and writing this book, and without their support this book would not have become a reality.

1   The Method of Inquiry

Introduction

The state of Texas has built a reputation for itself as the leading producer of crude oil and natural gas in the nation. Advancements in drilling technology like hydraulic fracturing or "fracking" in the state has only helped to further boost its oil and gas production and maintain its lead in energy production among the 50 states. Although fracking, like the electric car, may sound like a new technology to many, it has been present for several decades. It involves the use of horizontal drilling in conjunction with conventional vertical drilling to extract crude oil and gas from thousands of feet below in the sedimentary rocks. The attractiveness of this unconventional drilling practice lies in its ability to efficiently produce oil and gas from potential reserves in sedimentary rocks that were once considered inaccessible. This technology has made a timely comeback when crude oil and gas imports and prices were high in the nation and the federal government remained committed to its goal of energy independence. An added impetus was also a state climate that favored oil and gas drilling.

As fracking proved to be a success in Texas, the big and small oil and gas companies both in and outside the state lost no time in adopting this advanced drilling technology. The improved technology enabled the oil and gas industry to enhance their oil and natural gas production in the state. It prompted the companies to undertake massive exploration and drilling operations within the state and the investments paid off when massive sedimentary formations with potential reserves of oil and natural gas were identified within the state. With the proliferation of fracking within the state, it made gradual inroads into rural and urban communities.

During the period of fracking boom from 2009–2015, numerous rigs dotted the rural and urban landscape in the energy rich shale plays of the state. They served as a constant reminder that oil and natural gas production was taking place even amidst the regular activities of rural and urban life. The busy movements of onsite workers and the truck traffic plying back and forth to the production fields transformed the character of many quiet cities and towns into bustling centers of oil and gas production. The ancillary service industries including hotel, transportation, housing, food and others also received a boost while the state benefited from an increase in revenue from oil and gas production and sale of goods.

Research Focus

Behind the façade of economic gains loomed serious environmental concerns that failed to receive as much attention as the economic gains and rapid strides made toward energy independence. Residents living in rural and urban communities located close to drilling sites or where fracking was conducted within the boundaries of small towns and cities objected to its many negative externalities that were unforeseen at the time of adoption of this technology. It led to individuals' numerous complaints to the local and state governments that went unheeded. This added to their frustration and opposition to the drilling practice in many parts of the state. With the social outcomes of fracking not being as optimistic as the economic ones, this study has tried to investigate the regulatory and environmental aspects of fracking that have caused much discontent among people who have been adversely affected by this unconventional drilling activity.

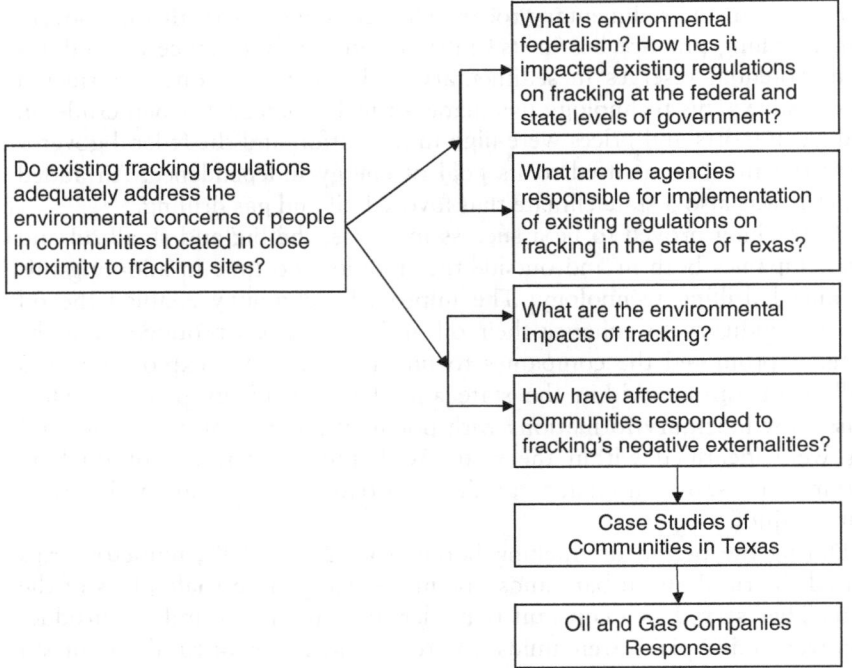

Figure 1.1 The Logic of Inquiry.

The research question that has been posed in the study is: "Do existing fracking regulations adequately address the environmental concerns of people in communities located in close proximity to fracking sites?" To answer this question calls for two things. First, investigation into the regulations on fracking at the federal and state level of government and the agencies responsible for their implementation. Second, a look at some of the environmental impacts of fracking and communities' responses to it.

The information will offer readers a glimpse into the existing regulations on fracking, create awareness of the environmental impacts of this advanced drilling procedure and enable readers to better understand the needs of individuals and the decisions of local and state governments at the juncture of economic gains and loss in environmental quality. Hopefully

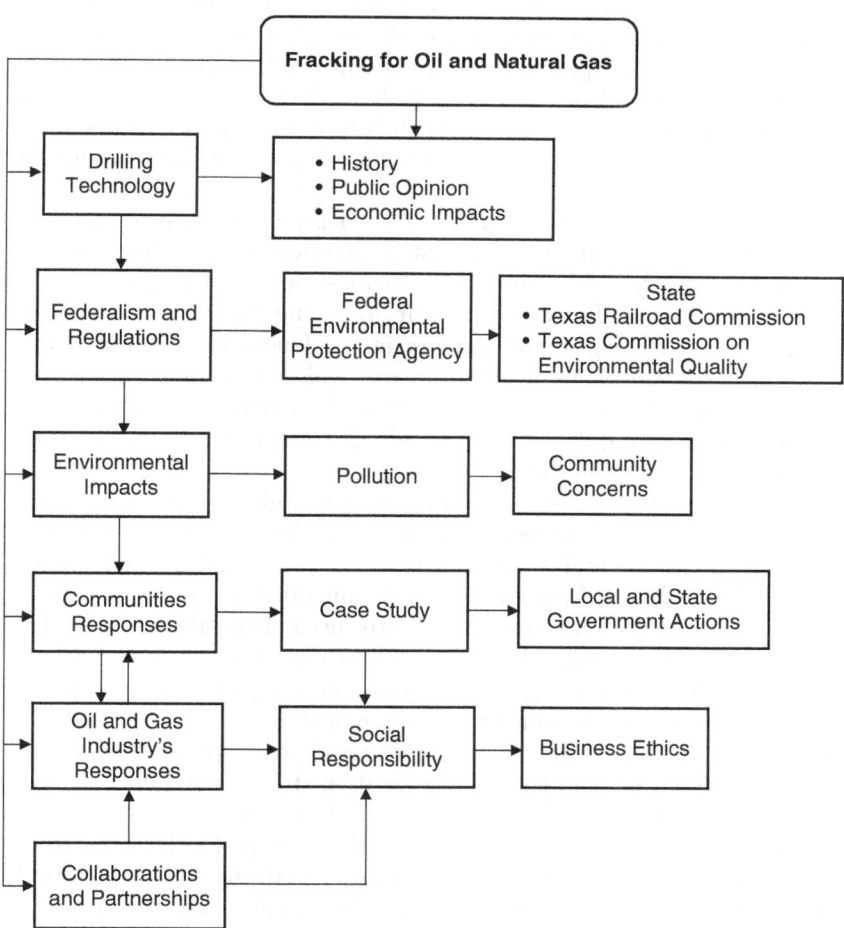

Figure 1.2 A Conceptual Framework of Fracking in Texas.

such information will help both individuals, decision makers, public administrators and the corporate world to agree on an efficient and effective strategy that will balance public interest with private ones without jeopardizing the prospects of goals of energy independence and expansion of geopolitical influence.

Sources of Information

In the collection and processing of information for the study, a qualitative research methodology has been used. Its selection can be validated on the grounds of versatility of its various research tools that have proven to be useful in the collection of both primary and secondary data. In the collection of primary data for the case studies, interviews were conducted. Individuals affected by fracking, community leaders, public officials, and other stakeholders were interviewed for the study. In selection of these people of interest, a snowballing approach was used where selection was made possible through a referral process. It helped to identify and include in the sample those individuals who had much information to share based on their personal experiences and or the role they played in community organizations.

The interviews were conducted over the telephone and participation in the study was on a voluntary basis. Each interview lasted approximately 40 minutes. Interviewees were asked various open-ended questions to probe the impacts of fracking on their lives and their involvement in actions, if any, to bring about changes in local policies. Their personal insights into the problem of fracking in local communities were also captured during the interview process. Overall, participants' inputs into the information seeking process simply added to the robustness of findings that have been reported in the case studies.

In the collection of secondary data, multiple sources were used. They included the websites of government agencies, the oil and gas industry, policy and environmental organizations, peer reviewed articles, newspapers, and even blogs. In retrieval of relevant information, content analysis and keyword searches were frequently used. Even the Google alert system proved very useful. It helped to collect the latest news on fracking that kept pouring into a designated mailbox on a regular basis and which was reviewed from time to time for reference purposes.

Further, in trying to offer a glimpse into residents' responses to fracking in communities close to fracking sites, a case study approach has been adopted because of the flexibility it affords in the collection and framing of information. As pointed out by Patton (2014), "a qualitative case study seeks to describe that unit in depth and detail, holistically, and in context." Additionally, a case study helps to present interesting facts and information from a past event or focus on a problem that is being confronted by a community. Whatever may be the issue or the problem, it is the unique aspect of each case that draws public attention and arouses their curiosity on the final outcome. Even various watchdog groups follow the progress of events in a case because of their personal or vested interest in it. The media also intervenes in the publicity of a case using both traditional (newspapers) and non-traditional (social media) modes of communication to evoke responses from people. Nowadays, even a random Google search on the internet on any topic can help to retrieve a plethora of information on various cases that offer noteworthy lessons. These lessons can be on success, failure, sustainability, and avoidance and can also be used as exercises in problem solving by academics or in decision making in various organizations.

Keeping in mind the essential elements of a case study, the cases presented in this book have focused on two things – (a) the externalities from fracking and their consequent impacts on the lives of those people who live at close proximity to drilling sites and (b) communities' responses to fracking. Through description and narration of details, each case has tried to draw attention to those events that are worthy of consideration in the context of bringing about relevant changes in the existing regulatory framework of the state and local governments.

Understanding Public Policy

We are all familiar with the words "public policy." All of us have heard endless times that we need to adhere to public policies, irrespective of our personal liking or dissatisfaction with them. Since public policies are inextricably linked to our lives, one needs to be reminded that their design and development is a complex process. Initially, it requires a political will to bring about a change. This is often followed by a chain of actions that require investment of time, resources, stamina, and perseverance to nurture a policy idea from its agenda to an implementation stage. In the journey of a policy idea from its inception to the action stage, its path is riddled with obstacles. The removal of obstacles calls for the involvement of several actors and each plays a unique role in the policy making process.

Focus on Regulatory Policy

Public policy can be classified into various types – regulatory, distributive, and redistributive (Lowi, 1964). Let us take the case of regulatory policy. Its impacts on the practice of fracking in urban and rural communities are considerable in the state of Texas and are the focus of the book. The very word "regulatory" is disliked by most people. Generally, people dislike any kind of restriction that is imposed on their activities. We hate to be told at what speed we should drive our car or where to dump our garbage. It is our independent spirit, disregard of the welfare of others, and selfish interest for profits that make us hate regulations. At the same time, from our innumerable experiences and failure of the market system, we have learned that governmental regulations can only protect us from the unfair trade practices that can deal a heavy blow to our lives and welfare.

Typically, regulatory policies are formulated and implemented to protect and serve the needs of the public and those who are regulated. Even though very few realize the dual purpose of such policies, they are often deemed as obstacles in our path and hated by those whose activities are regulated. Regulations are used by the government as a means to curb the unfair practices of those individuals and companies that render harm to the public and put their lives and welfare at risk. Violation of regulations can be costly, making companies pay millions of dollars as fines and compensations to the victims of unfair trade practices.

The social benefits of regulation far outweigh the economic costs involved in compliance of government-imposed standards. For example, if not for regulations in the Clean Air Act, flaring of waste chemicals like volatile organic compounds such as sulfur dioxide, waste gases, and benzene, a well-known carcinogen, would have remained unchecked from a refinery and chemical plant owned by Shell and its affiliated partnerships. This would have exacerbated toxic air pollution in the area and inflicted serious health damages on people in the surrounding community of Deer Park area near Houston. For the alleged violations of the Clean Air Act in 2013, the company settled with the U.S. Department of Justice and the Environmental Protection Agency to spend at least $115 million on upgrading its facility to curb toxic air pollution and pay $2.6 million as civil penalty (Department of Justice, 2013). Another example that helps to reiterate the importance of regulations is the case of Volkswagen cheating on emission testing in the U.S. The company had installed software to cheat on emission testing in 11 million vehicles that were sold in the U.S. (Gates, Ewing, Russell & Watkins, 2017). If not for existing regulations on emission testing, such a trap would have gone undetected. The Volkswagen cars would have continued to spew tons of harmful pollutants into the air from tailpipes and wiped off some of the gains made in air quality as a result of strict enforcement of emission standards on automobiles.

Regulations have delivered many benefits to the public and the industry and do cost money to both when implemented. The public as taxpayers have to fund the agencies responsible for implementation while the companies have to pay for retro-fittings and investment in new technology that are required by regulations. Many times regulations have saved the public from becoming a victim of unfair trade practices and paying heavily in terms of money or lives and sometimes both as seen in the case of faulty Takata air bags that were installed in Honda, Ford, and other auto manufacturers' vehicles. When it comes to the environment, even though the public relies more on regulations than on the market to make a transition to renewable energy in the twenty-first century, there seems to be an even split on the issue of whether fewer regulations can protect air and water (Funk & Kennedy, 2017).

In a free enterprise economy, just as too many regulations can be oppressive and are considered unconducive for industrial growth, laxity in regulation and deregulation can be equally worrisome. For example, small oil and gas companies in West Texas availed an exemption that allowed sources that emit less than 25 tons of sulfur dioxide and volatile organic compounds to be exempted from the strict permitting requirements of the Clean Air Act. As per the exemption, these companies were not required to issue public notices and invest in advanced technology to abate air pollution. Unfortunately, the exemption granted in good faith was violated when oilfields released ten million pounds of pollutants into the air in 2016, which posed a serious threat to public health in the surrounding community (Collett, 2017).

Even efforts to minimize regulations and promises of self-policing may spell victory for the oil and gas industry and be well worth the lobbying expenses, but in the long run they do the public little good. People's lives and welfare are at risk when there is no way for them to find out about hidden dangers lurking close by. In such circumstances, governmental regulations can provide the needed protection to the public, provided politicians do not succumb to the pressures of the industry for deregulation. The latter is evident under the Trump administration. In response to oil and gas industry lobbyists' demand for the removal of burdensome regulations, the federal government has relaxed rules on methane emissions from fracking sites. Contrary to expectations, this deregulation has not been greeted with equal relief from all oil and gas companies. Despite the ongoing skepticism on climate change, some of the larger oil companies like Exxon have come to terms with the idea of human induced climate change from the combustion of fossil fuel. They now support some of the Obama era rules that the Trump administration is trying to repeal. For example, use of a carbon tax, capture of methane from drilling sites to reduce emissions, and reporting of data on emissions. The adoption of such a stance comes with these companies' expectations of an early energy transition by 2020 and the realization that government policies can help to make the transition to low carbon energy more competitive (Tomlinson, 2018).

Policy Terms

In the book, in discussions on regulations on fracking and in trying to understand the regulatory stance of the government, several concepts have been borrowed from the literature on public policy analysis along with those from the field of public administration. For example, the term "window of opportunity" (Kingdon, 1995) is commonly used in policy literature to refer to the opportune moment for introducing or making amendments in a public policy. During the Obama administration, the window of opportunity to make changes in environmental regulations was open. Environmentalists and policy activists availed the opportunity to make changes in existing environmental regulations and introduce new ones. For example, a policy on climate change was introduced in the best interests of the United States and global citizens. From 2017 onwards under the Trump administration, the window of opportunity has closed. Regulations imposed during the Obama era have been revoked in the name of efficiency, cost savings, and creation of jobs, without taking into consideration the long-term consequences of such actions. For example, the construction of the North Dakota pipeline that was stalled during the Obama administration has been given the order to proceed under the expectations that it will help to create jobs and support energy development from the fracking of Bakken Shale deposits in North Dakota (Eilperin & Dennis, 2017). In describing the interrelationships between different levels of government and their respective regulatory authority, the concept of "environmental federalism" has been used to explain the distribution of regulatory authority between the federal and state governments.

Another word used in the book that is of interest in the policy and public administration literature is "capture." Used several times in the book, it refers to the influence exerted by lobbying groups on the regulatory decisions and actions of government agencies. Even the existing statistics on money spent by the oil and gas industry to lobby the government agencies hint at a capture event and the influence of the industry on both state politics and its regulatory arm of policymaking. Other terms like "network" and "collaboration" have also been used to better understand the process of group dynamics that attempt to bring about changes in policies at all levels of government. For example, citizen groups against fracking in communities often network and form alliances with each other as they continue to wage their fight against fracking's externalities. It is collaboration that enables citizens' groups and large non-profit environmental

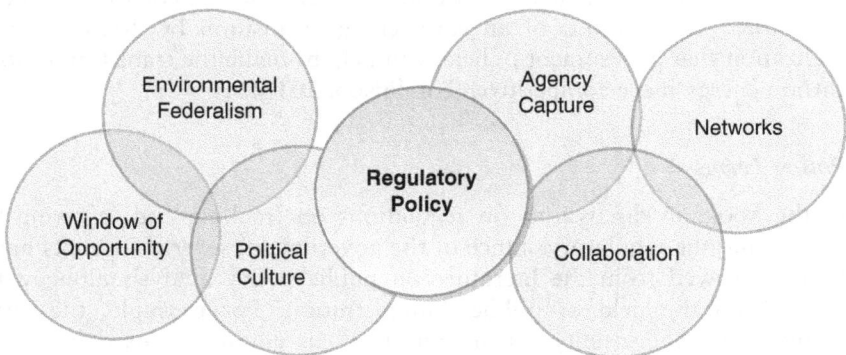

Figure 1.3 Interrelated Concepts in Public Policy Implementation and Administration.

organizations to rally against a common cause and put pressure on government to impose restrictions on industries' activities so that they do not undermine public welfare.

Political Culture

The concept of political culture has been applied to understanding the political foundation of state regulations. No single definition of culture has been able to capture the dynamics of this concept. As a result, scholars in political science and public policy have made various attempts to convey its actual meaning. Some consider it the "distribution of patterns of orientations towards policy action" (Almond & Verba, 2015). Others regard it as a "mindset" with a favorable disposition toward a range of alternatives (Elkins & Simeon, 1979) or a publicly expected behavior that may not be preferred individually (Chilton, 1988). Irrespective of the suggestions, the importance of political culture lies in the fact that it has a greater impact on the behavior of political leaders than on public opinions (Boeckelman, 1991).

In the study, Elazar's (1972) ideas on political culture of states have been applied to understand the influence of culture on the regulatory behavior of states. In his book *American Federalism: A View from the States*, Elazar has identified political subcultures that coexist in the 50 states. Even though this book was written decades ago when the Hispanic culture was not prominent, the author's ideas on political subculture are still somewhat relevant and can be applied to understand states' behavior when it comes to policy and other related issues.

According to Elazar, there exist three types of political subcultures in the U.S. They include the traditionalists, individualists, and moralists. In a traditional political culture, the public plays a passive role in state politics while the elites with wealth and power tend to exert a greater control on state economy and in maintaining the social status quo. While in an individualist culture, there is more reliance on the marketplace to serve the interests of the people. In this culture, politicians run for office for personal and economic gains. It is the moralist culture that is noted for its consensus among both people and politicians in the attainment of a common public good and pursuing politics to serve the interests of the people.

There exists a combination of subcultures in a state which define the political makeover of the nation. In a state, two or more subcultures may coexist. For example, in the state of Texas, the two predominant sub cultures include those of traditionalist and individualist. While in the state of New York, moralist and individualist subcultures are the two prominent ones. Each state with its distinct political subcultures strives toward a goal agreed upon by a vast majority of people belonging to both the subcultures.

Theories and Framework

Against a background for the need of a regulatory policy, efforts have been made in this book to understand the rationale used in the decision-making process by decision makers and the attempts made by community residents to advance their problems to the agenda stage of policy making with the expectation of a favorable policy outcome. Since no single theory or framework can offer a full understanding of a policy change or lack of it, attempts have been made in this book to analyze various policy related events and information through the lens of selected theories and frameworks. They include the Prospect Theory, Multiple Streams Approach, and the Advocacy Coalition Framework. Although each have their own limitations, they still play an important role in understanding various policy related events by focusing on various aspects that decide policy outcomes. For instance, the focus is on human behavior in Prospect Theory, in the Multiple Streams Approach it is on social, economic, institutional values and interests that compete with each other to bring prominence to a problem, while the focus is on coalitions in the Advocacy Coalition Framework. Each has been briefly described in the following sections.

Prospect Theory

Prospect Theory is an economic theory that is often applied to understand decision making under conditions of risk. Unlike other economic theories, where individuals are portrayed as self-interested rational beings, this theory recognizes the importance of psychological and social aspects of human behavior in their decision-making processes. It takes into consideration norms, habits, and expectations of people that play an important role in decision making. Developed by Kahneman and Tversky (1979), the theory helps to understand how individuals make decisions through evaluation of losses and gains from a mental anchor or a reference point. This reference point is usually the status quo. Depending upon how a change is framed, that is either as a gain or a loss, individuals are likely to make a decision from this reference point.

As per the tenets of this theory, individuals tend to attach more importance to losses than to gains. That is, individuals are risk averse to gains and risk acceptant toward losses. This type of behavior can be attributed to the economic fact that the marginal utility from gains diminishes faster than that from the marginal disutility of losses. Such a phenomenon leads individuals, either as a single person or as a group, to value more what they have than what they do not have. For example, it is the fear of loss from malicious cyber-attacks that make individuals spend hundreds of dollars on purchases of protective software services when that money could have been well spent on various other productive or gainful pursuits.

Additionally, the theory posits that sometimes the act of overvaluation or the endowment effect impacts on a person's sense of judgment of fairness and justice in decision making (Kahneman & Thaler, 1991). There are other factors that play important roles in individuals' decision making which include pseudocertainty and isolation effects. As the word implies, pseudocertainty sometimes makes individuals overweigh small probabilities and underweigh the smaller and moderate ones. This leads to the treatment of uncertain outcomes as if they were certain in the decision-making process. On the other hand, the isolation effect leads to the disregard of common elements among the alternative options and to a focus on a component that is different. Individuals often edit their choices under the influence of these effects. The editing process is often followed by evaluation of prospects where prospects with higher values are usually chosen over the smaller ones.

Multiple Stream Approach

When there exist several problems in our society with each vying for attention, it becomes difficult sometimes for elected officials and others to prioritize a problem and seek a policy solution. The Multiple Streams Approach of John Kingdon (1994) offers some insights into the processes involved in the advancement of a problem to the agenda stage of the policy cycle. Though obstacles like financial constraints, opposition by powerful interest groups, lack of public attention, politics, and weak leadership often impede the progress of a problem to the agenda platform, they do not discourage people from taking the necessary actions to seek a politically expedient solution to their problem. They invest their time and resources to advance their problem to the agenda stage and also propose technologically and politically feasible options or alternatives for consideration in solution of the problem.

The Multiple Streams Approach is based on the ideas of the Garbage Can model of Cohen, March and Olsen (1972). In the garbage can model, the focus is on "organized anarchies" where there exist four streams – problems, solutions, participants, and choice opportunities. Each stream has a life of its own and is unrelated to the others. Kingdon (1995) revised the model to include three separate streams: problems, policies, and politics. These streams develop separately and are controlled by different forces, considerations, and styles but are not totally independent of each other.

In a problem stream, there exists a distinct problem that has been identified by people either through indicators, dramatic events, or feedbacks from existing programs. A policy stream refers to a short list of ideas that have been agreed upon by various categories of people in a community who are keen on bringing about policy changes. The political stream is composed of multiple elements, which include public mood, election results, ideology changes of decision makers, administrative changes, and pressure from interest group campaigners. When the three streams are in alignment, the window of opportunity opens up. This policy window remains open only for a short time and policy entrepreneurs must seize the opportunity to couple or join the streams at the critical moment to bring about a policy change. For instance, during the Obama administration, environmental issues which failed to receive adequate attention and garner support for policy action during the Bush era received much attention. This prompted the environmental groups to push their agenda through the policy window. They played an active role during this short period and were able to influence the federal government's decision to stop the construction of the huge Keystone XL pipeline, from Canada to Nebraska, in late 2015 and restrict the emissions of methane from oil and gas wells in federal and tribal lands. In early 2017, two days after President Trump took office, the policy window closed. He signed an executive order to reverse the Obama era decisions. By signing a memorandum, he has made it possible for TransCanada, the pipeline construction company, to re-apply for the permit to construct and operate the pipeline. But there is a caveat: the pipes have to be manufactured in the United States (Daly & Thomas, 2017). Also, under the Trump administration, the Environmental Protection Agency has delayed compliance with Obama era rules that limited methane emissions from oil and gas drilling sites (Eilperin, 2017). The quick passage and derailing of President Obama's legacy reminds us that the window of opportunity can quickly open and close. If a cycle is missed, policy actors need to wait for the next critical and opportune moment to arrive.

Keeping these facts in mind, the multiple stream approach of Kingdon (1994) has been applied in the study to scrutinize individuals' efforts to draw attention to the problems emanating from fracking related externalities in their communities, with the expectation of a favorable policy outcome.

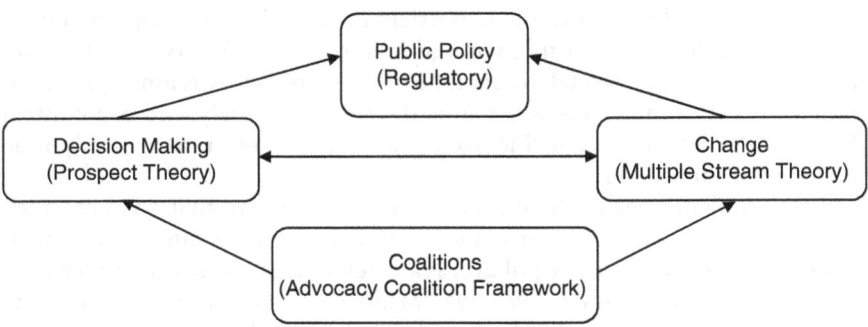

Figure 1.4 Tools Used to Analyze Policy Related Information.

Table 1.1 Essential Elements of Theories and Framework

Framework	Author	Application and Important Characteristics
Prospect Theory	Kahneman and Tversky (1979)	In decision making • Estimation of loss or gain from a reference point or the status quo • Endowment effect makes individuals risk averse to gains and risk acceptant to losses • Psuedocertainty or tendency to overweigh small probabilities and underweigh high to moderate ones • Isolation effects lead to disregard of common elements and focus on a component that is different • Editing of options to facilitate choice • Evaluation of prospects leading to selection of those with higher values
Multiple Stream Approach	John W. Kingdon (1994)	For an issue to reach the agenda setting stage • Problem identification • Making policy proposals • Political support • Opening of window of opportunity • Advocacy by policy entrepreneur
Advocacy Coalition Framework	Paul Sabatier (1988)	For a policy change • Time span of decade or more • Existence of policy subsystems • Intergovernmental dimensions • Belief systems • Policy oriented learning

Advocacy Coalition Framework

Further, to understand the role played by coalitions in bringing about policy changes at a local level of government, the Advocacy Coalition Framework (ACF) of Paul Sabatier (1988) has been selected in the study. According to the ACF, policy changes and policy-oriented learning typically become evident over a period of a decade even though initial efforts may lead to its failure. Since the issue under consideration in the study does not fulfill the time criterion for a policy change, the ACF model has only been applied partially. That is, ideas on coalition formations and group dynamics have been borrowed from ACF to understand the scope of policy innovation at the local and state level of government in the state of Texas.

In the ACF model, the structural components include policy subsystems, the intergovernmental dimension, and belief systems. They are regarded as essential in bringing about a policy change. Among the components, the policy subsystem is composed of actors from public, private, and non-profit organizations who share a similar concern and goal aimed at promoting individual welfare. The intergovernmental dimension refers to the discretionary power of local and state government officials in bringing about a policy change. Lastly, the belief systems allude to value priorities, perceptions of causal relationships and evaluations of existing policy mechanisms by those individuals who are actively engaged in bringing about relevant policy changes in communities that are negatively impacted by existing policy directives.

To explain those events where the ACF model cannot be applied, efforts have been made to identify those structural components that serve either as drivers or non-drivers of policy change at the local level of government. One such component is culture. According to the cultural theory, "culture is a system of persons holding one another mutually accountable" and "each type of culture is based on a distinctive attitude towards knowledge" (Douglas, 1993). This theory upholds four worldviews of people in society: the *individualists*, who tend to be more risk acceptant in a domain of economic gains; the *egalitarians*, who are willing to forego economic gains to reduce their level of risks; the *hierarchicalists*, who dislike any kind of threats arising from the social order; and the *fatalists*, who live in a world of constant fear with the realization that they have little or no ability to control their own world. The cultural theory has been previously used to explain people's attitude toward the environment. Often the egalitarian and individualistic values of individuals play an important role in eliciting responses to environmental risks when they perceive a likely change in it (Carlisle & Smith, 2005). To what extent those values are evoked depends upon individuals' involvement in the society. When involvement or interactions with others in society is weak, it is the individualistic attitude that prevails while the opposite is true when group involvement is intense. The interactions in a group often lead to the exertion of a subtle pressure on individuals to conform to the group's norms and decisions in the best interests of all those experiencing a problem in the society.

Conclusion

The qualitative methodology applied in the study has helped to interpret and understand the actions of government agencies, private industry, and of those citizens who live in communities that have undergone remarkable changes due to fracking. Its various tools have helped to understand the rationale in the decision-making process, policy implementation, and the efforts made by coalitions to bring about relevant policy changes. The various events and regulations discussed in the course of the study help to draw attention to individuals' and industries' efforts in both regulation and deregulation of public policies to promote their interests. All such events remind us of the overall complexity in the formulation and implementation of regulatory policies and the consequences for the economy and society.

By filtering the various events through the lenses of theories and a model, it has been possible to throw light on the continuous tug of war that exists between public and private interests in the regulatory policy world. It is the winner who gets to influence the direction of a regulatory policy, irrespective of the time frame. In the following chapters, a critical look at the regulations and their implementation in a politically charged climate helps us to understand the efficacy of regulations in a free enterprise economy and thereby answer the research question posed in the study. The theories serve as a constant reminder that individuals' sense of losses, both monetary and in environmental quality, serve as the impetus to seek policy actions. The efforts made in such a pursuit are equally complex. They call for organization, leadership, and investment of resources and time to facilitate the coupling of problems to policies and politics. Even then, such actions can be pursued only at opportune moments in order to reap the benefits of anticipated outcomes.

Further, the case study approach used in the book has offered an understanding of the role played by various groups of people in bringing about a policy change. The model used to understand group dynamics has made obvious two things. First, groups have to surmount many hurdles prior to negotiations, bargaining, and collaboration and, second, unity among groups is essential in bringing about a desired policy outcome in a community.

References

Almond, G. A. & Verba, S. (2015). *The civic culture: Political attitudes and democracy in five nations.* New Jersey, Princeton University Press.

Boeckelman, K. (1991). Political culture and state development policy. *Publius,* 49–62.

Carlisle, J. & Smith, E. (2005). Postmaterialism vs. Egalitarianism as predictors of Energy-related attitudes. *Environmental Politics, 14*(4), 527–540.

Chilton, S. (1988). Defining political culture. *Western Political Quarterly, 41*(3), 419–445.

Cohen, M., March, J. & Olsen, J. (1972). A garbage can model of organizational choice. *Administrative Science Quarterly, 17*(1), 1–25.

Collette, M. (2017, July 7). Report: State lets polluters off hook. *Houston Chronicle,* A1 and A3.

Daly, M. & Thomas, K. (2017, January 24). Trump acts to advance Keystone XL, Dakota Access pipelines. Associate Press. Retrieved from www.yahoo.com/news/trump-acts-advance-keystone-xl-dakota-access-pipelines-172210015-finance.html.

Department of Justice. (2013). Shell oil to spend over $115 million to reduce harmful air pollution at Houston area refinery and chemical plant. Retrieved from   www.justice.gov/opa/pr/shell-oil-spend-over-115-million-reduce-harmful-air-pollution-houston-area-refinery-and.

Douglas, M. (1992). *Risk and blame: Essays in Cultural Theory.* Taylor and Francis, New York.

Eilperin, J. (2017, June 14). Trump administration delays rules limiting methane emissions. *Washington Post.* Retrieved from www.washingtonpost.com/politics/trump-administration-delays-rules-limiting-methane-emissions/2017/06/14/0e7d50fa-512b-11e7-be25-3a519335381c_story.html?utm_term=.660879ae283a.

Eilperin, J. & Dennis, B. (2017, February 7). Trump administration to approve final permit for Dakota access pipeline. *Washington Post,* February 7, 2017. Retrieved from www.washingtonpost.com/news/energy-environment/wp/2017/02/07/trump-administration-to-approve-final-permit-for-dakota-access-pipeline/?utm_term=.7db8013fbea6.

Elazar, D. (1972). *American federalism: A view from the states* (3rd ed.). New York, Harper Collins.

Elkins, D. J. & Simeon, R. (1979). A cause in search of its effect, or what does political culture explain? *Comparative Politics, 11*(2), 127–145.

Funk, C. & Kennedy, B. (2017). Public divides over environmental regulations and energy policy. Pew Research Report, May 16, 2017.

Gates, G., Ewing, J., Russell, C., & Watkins, D. (2017, March 16). How Volkswagen "defeat devices" worked. *New York Times.* Retrieved from www.nytimes.com/interactive/2015/business/international/vw-diesel-emissions-scandal-explained.html?_r=0.

Kahneman, D. & Tversky, A. (1979). Prospect theory: An analysis of decision under risk. *Econometrica: Journal of the Econometric Society,* 263–291.

Kahneman, D., Knetsch, J., & Thaler, R. (1991). Anomalies: The endowment effect, loss aversion, and status quo bias. *The Journal of Economic Perspectives, 5*(1), 193–206.

Kingdon, J. W. (1994). Agendas, ideas, and policy change. *New perspectives on American politics,* 215–229.

Kingdon, J. W. (1995). *Agendas, alternatives and public policies* (2nd ed.). New York, Longman.

Lowi, T. J. (1964). *American Business, Public Policies, Case-Studies, and Political Theory, 16*(4), 677–715.

Patton, M. Q. (2014). *Qualitative evaluation and research methods, fourth edition.* SAGE Publications.

Sabatier, P. A. (1988). An advocacy coalition framework of policy change and the role of policy-oriented learning therein. *Policy Sciences, 21*(2), 129–168.

Tomlinson, C. (2018, February 11). Oil giants are ready for action on climate. *Houston Chronicle,* B1 and B9.

2 Fracking for Oil and Natural Gas in Texas and its Impacts on the Economy

Introduction

Hydraulic fracturing has revolutionized the extraction process of crude oil and natural gas from unconventional sources like shale and coalbed methane rocks that were once difficult to access. Now that the technology has been improved and is yielding a large supply of natural gas and some crude oil, it has been partly blamed for the recent glut in oil supply in the global market and fall in natural gas prices in the United States market. To better understand the process of hydraulic fracturing, one needs a brief description and understanding of its historical origin. Technologically, hydraulic fracturing can be defined as a process or step in well drilling to extract oil and gas from sedimentary rocks lying deep below at depths ranging from 1,500 m to 3,000 m. It is often used in conjunction with horizontal drilling. Hydraulic fracturing helps to create the pathways for the escape of oil and gas from sedimentary rocks that were once considered out of reach and therefore not economical for drilling. Also known as "fracing" or "fracking," hydraulic fracturing is not an entirely new technique but instead has been improved over the years with advancements in technology and the need for cheap energy in a nation's quest for energy independence. The word "fracking" will be used in the rest of the chapter and throughout the book to refer to hydraulic fracturing.

Fracking

The initial development of well simulation, which later came to be called "hydraulic fracturing," dates back to the 1860s when gun powder was initially used to detonate and create large holes to extract oil and gas from rock formations. In the twentieth century, the technology underwent a series of innovations and experimentations followed by a major change in the 1990s, enabling the commercial production of natural gas and crude oil from shale formations in a cost-efficient way. The modern process of fracking involves the injection of water, sand, and chemical additives under high pressure through horizontally drilled wells to create multi-stage fractures and enlarge existing ones in shale rock formations. The fractures and crevices facilitate the migration of crude oil and natural gas from once hard to reach oil and gas reserves in shale rocks to the existing wells, from where they are extracted and brought to the surface. A single well can be fracked several times and fracking helps to increase production anywhere from 5 to 15 percent. However, fracking is a water intensive process. It requires large quantities of water and sand ranging from 45,000 gallons of water and 35 tons of sand for a small well to several million gallons of water and 1,500 tons of sand for a large well. For example, in the fracking operations of each production well in the Marcellus Shale of the Appalachians (Susquehanna River Basin Commission, 2011), water consumption ranges from four to five million gallons while in the Eagle Ford Shale of Texas, it is 13 million gallons (Nicot et al., 2011). Also, the chemical additives used in fracking tend to vary with the rock structure and the type of product to be extracted (Hilyard, 2012). The oil and gas companies possess proprietary rights over their use of mixture of chemicals in drilling for oil and gas. Although some states do not require the public disclosure of the names of all toxic chemicals used, they do need to keep the state regulatory agencies informed.

The development and innovation of the fracking technology in the United States can be attributed to several factors. Among them, governmental factors seem to emerge as important role players both directly and indirectly at the federal and state levels. Usually, the federal government in its quest for energy independence invests millions of dollars to develop new energy resources like nuclear, wind, solar, battery technology, hydrogen fuel cells, and others. In the development and production of fossil fuel, the federal government has failed to display the same level of enthusiasm and investment as observed in other forms of energy development. Nevertheless, the federal government offered a tax break through a 1980 tax bill (section 29) to the oil and gas industry. As per the provisions of this bill, oil and gas companies were promised a special federal tax credit for the development of technology to harness unconventional natural gas. This bill provided the impetus to big and small companies to experiment with the fracking technology in drilling for the unconventional gas.

In the race for extraction of the unconventional gas and to avail the tax break, George Mitchell, the owner of a small energy company in Texas, started drilling in the Barnett Shale play of northern Texas. Here, he stumbled upon the right combination of horizontal drilling and use of pressure and mixture of fluids to pry open the fractures in the sedimentary rocks to release the trapped natural gas. Using a trial and error process, Mitchell was successful in bringing about improvements in the fracking technology without resorting to heavy-duty research in energy laboratories. As we look back, it is evident that the federal government's incentive and Mitchell's spirit of entrepreneurialism made it possible for the United States to increase oil and gas production and make progress toward the nation's goal of energy independence (Gertner, 2013).

Over the years, other factors have also helped to promote the fracking technology after it attained a cost-efficient status. For example, oil and gas companies do not need to seek the state government's permission for fracking in privately owned lands beneath which lie reserves of oil and natural gas. The private companies can enter into lease agreements with property owners at their own discretion in return for royalty payments. It is only in government owned or public lands that permission is required for fracking. Also, the existing infrastructure of oil and gas pipelines has enabled the oil and gas companies to venture into the business of fracking. The interstate pipeline infrastructure has made it possible for both big and small oil and gas companies to transport their end products to the refineries and markets dispersed throughout the nation (Merrill, 2013).

At the state level, the development of agencies to control oil and gas activities in the state has benefited the exploration and development of oil and gas drilling activities using the fracking technology. The prospect of economic benefits likely to be reaped through energy exploration and production has helped most states to develop a favorable stance toward fracking and a lackadaisical attitude toward regulations. States continue their domination in the regulation of the oil and gas industry as the federal government only entered the regulatory arena with its environmental laws much later in the 1970s (Merrill, 2013). States do not typically like federal intervention in state matters of environmental protection, but under the concept of environmental federalism or sharing of power between the federal and state governments, states are required to follow the guidelines developed by the Environmental Protection Agency (EPA) in implementation of federal regulations over the oil and gas industry. But the Halliburton Loophole in the Energy Policy Act of 2005 has exempted fracking from regulation under the Clean Water Act, Clean Air Act, and other environmental statutes.

With fracking gaining momentum in various parts of Texas and other states in the nation, individuals living in communities close to fracking sites have experienced first-hand both the ills and benefits of this oil and gas extraction process. Based on the nature of their experiences and gradual familiarity with this type of drilling procedure, they have either supported or opposed it. Whatever may be the public opinion, the oil and gas rich states have continued to lend much support to the oil and gas industry's fracking operations. An exception to this can be observed in the states of New York and Florida where strong public sentiment against fracking has discouraged its practice despite its many economic benefits.

The Public's Attitude toward Fracking

To gauge public sentiment on fracking, various surveys have been conducted throughout the nation over the last few years. Starting with the energy production process, a 2016 Gallup poll has revealed that more people (73 percent) favor the production of energy from alternative sources like wind and solar than from oil and gas (21 percent). Despite the high level of bipartisan support for renewable energy and public dissatisfaction with current energy policy, the federal and state governments continue to support fracking on public and privately leased land.

A closer look at the map of the United States reveals that notable shale formations credited with large reserves of oil and gas include the Marcellus Shale in the northeast, Barnett Shale in the southwest, Bakken Shale in the north, and the Colorado group in the west. See Figure 2.1. The federal government has long recognized the importance of fracking in the nation's quest for energy independence and security. Starting with the passage of the Mineral Leasing Act of 1920 to the current Trump

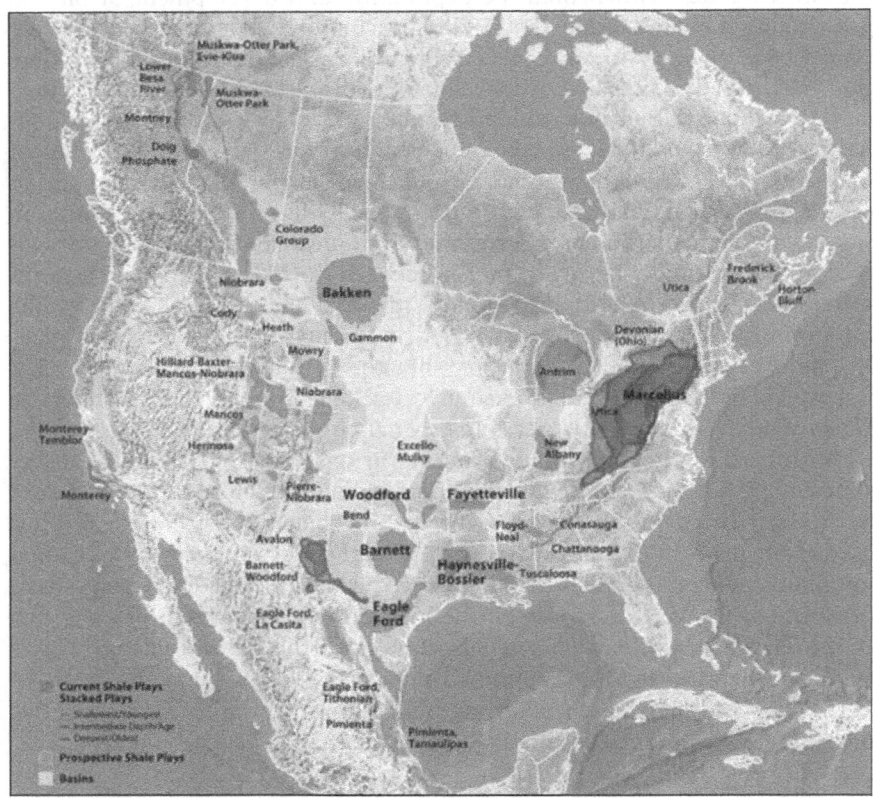

Figure 2.1 Shale Plays of the U.S.A.

Source: Energy Information Administration. Available at: www.eia.gov/oil_gas/rpd/northamer_gas.pdf.

administration, oil and gas drilling and later fracking have always had strong support from all levels of government. Even in most state governments, fracking is viewed favorably. It has helped to create jobs and generate considerable revenue for the local and state governments and contribute to the nation's Gross Domestic Product. During the period of 2011 to 2014, the fracking boom helped to create a large number of jobs in both skilled and unskilled categories and provided a major economic boost, particularly to the states of North Dakota and Texas. Despite the many economic benefits of fracking, it has faced much opposition at the local level of government from those people who have been negatively impacted upon by its externalities.

Over the years, the increased adoption of this technology in the extraction of oil and gas has helped to create a knowledge base on both the positive and negative aspects of fracking. At the beginning of the twenty-first century, such information was non-existent. The media has been quick to report on the attributes of fracking and has helped to increase awareness and shape public opinion on the issue. According to a 2012 Pew study, only 26 percent of people were aware of fracking, 37 percent had heard little about it, and 37 percent had no knowledge of it. In 2015, another study by the Center for Local, State and Urban Economy reported an improvement in the numbers – 41 percent of the population was aware of fracking, 36 percent had heard little about it, and 18 percent had no knowledge of it. Their level of awareness also varied with their educational attainments. While only 10 percent of individuals with high school education had heard a lot about fracking, 40 percent of those with a graduate or a professional degree had heard a lot about fracking (Borrick & Clarke, 2016).

Among those who had heard about fracking, their political affiliations impacted on either their support or opposition to it. According to the 2012 Pew study, 73 percent of Republicans favored fracking, 15 percent opposed it, and 12 percent did not know about this technology at all. Among the Democrats, only 33 percent favored fracking, 52 percent opposed it, and 15 percent did not know about it. A follow up study on fracking by the Pew Research Center in 2015 has revealed that even though Republicans continue to favor fracking, there has been an overall decline in their support – the proportion of Republicans supporting fracking was 57 percent in 2015. Conversely, opposition to fracking among Republicans increased to 33 percent in 2015. During the same time period, support for fracking among Democrats declined slightly to 30 percent while opposition to it increased to 62 percent.

Socio-economic factors also play an important role in influencing people's opinion on fracking. Supporters of fracking tend to be older, possess a bachelor's degree, watch television news, are politically conservative, and are well aware of both the economic and energy impacts of fracking. On the other hand, opponents of fracking tend to be women with greater

Table 2.1 Individuals' Attitude toward Fracking (%)

Polling Agency	Survey Year	General Support	General Opposition	Partisan Support	Partisan Opposition
Pew	March 2012	52	35	Republicans – 73 Democrats – 33	Republicans – 15 Democrats – 52
Pew	March 2013	48	38	Republicans – 48 Democrats – 38	Republicans – 66 Democrats – 33
Pew	November 2014	41	47	Republicans – 62 Democrats – 29	Republicans – 25 Democrats – 59
Pew	August 2014	39	51	Republicans – 57 Democrats – 30	Republicans – 33 Democrats – 62
UT Austin	January 2016	47	37	Republicans – 65 Democrats – 30	Republicans – 37 Democrats – 54
Gallup	March 2015	40	40	Republicans – 66 Democrats – 26	Republicans – 20 Democrats – 54
Center for Local, State and Urban Policy	September 2015	35	39	Republicans – 53 Democrats – 30	Republicans – 23 Democrats – 49
Gallup	March 2016	36	51	Republicans – 55 Democrats – 25	Republicans – DU* Democrats – DU

Source: Based on data from several polling agencies' polls.

Note
* DU – Data Unavailable.

awareness of the environmental impacts of fracking, read newspapers regularly, and entertain ideas of equality of rights and opportunities (Boudet et al., 2014).

People's opinions on fracking not only waver with time but also differ based on their geographical locations. A 2014 Pew survey (Pew Research Center, 2015) has found that more people in the northeast (59 percent) and the west (58 percent) oppose fracking in comparison with those who live in the midwest (46 percent) and the south (44 percent). Further, data collected by various polls have shown that public support for fracking is largely influenced by its environmental impacts and the current prices of oil and natural gas.

In 2012, when the price of a barrel of crude oil was slightly over \$110, fracking enjoyed the highest support. According to a 2012 Pew survey, 52 percent of the population supported fracking as they believed in its potential to achieve energy independence. In 2016, the sharp drop in oil prices to less than \$40 a barrel and greater awareness among people of the environmental and health impacts of fracking have made many withdraw their support for it. A 2016 Gallup poll revealed that only 36 percent supported fracking while 51 percent opposed it.

With the abundance and availability of cheap fossil fuels and concurrent increase in renewable energy production, a large portion of the population did not feel the need for fracking to supplement the nation's supply of energy. Their environmental values also overrode the influence of "not in my backyard" or the NIMBY syndrome in their opposition to fracking in the community (Michaud et al., 2008).

The influence of the media on the public's perception of fracking cannot be overlooked. The polarity in the news published, on both the positive and negative aspects of fracking, have left many people confused. For example, elite newspapers like the *Washington Post* have focused on the economic benefits of fracking while the *New York Times* published the maximum number of articles on fracking (70 percent). Both the *New York Times* and the *Los Angeles Times* have mainly reported on negative aspects of fracking on the environment, climate change, and water pollution along with discussions on responsibility and blame (Habib & Hinojosa, 2016). From the conflicting news and people's opinions on fracking, it is evident that if fracking is still going to be continued for political and economic reasons, more individuals would like to see this activity subjected to stricter regulations to reduce the negative impacts on the environment and public health (Davis & Fisk, 2014).

Fracking and Fossil Fuel Production

Whatever may be the public opinion on fracking, advancements in the technology have certainly made it possible to tap into huge reserves of natural gas and crude oil in various shale formations of the United States and expand domestic production. Fracking has contributed to an increase in the domestic production of natural gas by 35 percent during the period of 2008 to 2015, which has helped to reduce natural gas imports by 31 percent during the same time period. As a result of fracking, there is now an abundance of natural gas supply in the domestic market. The availability of natural gas at a low price has enabled the utility industry to make

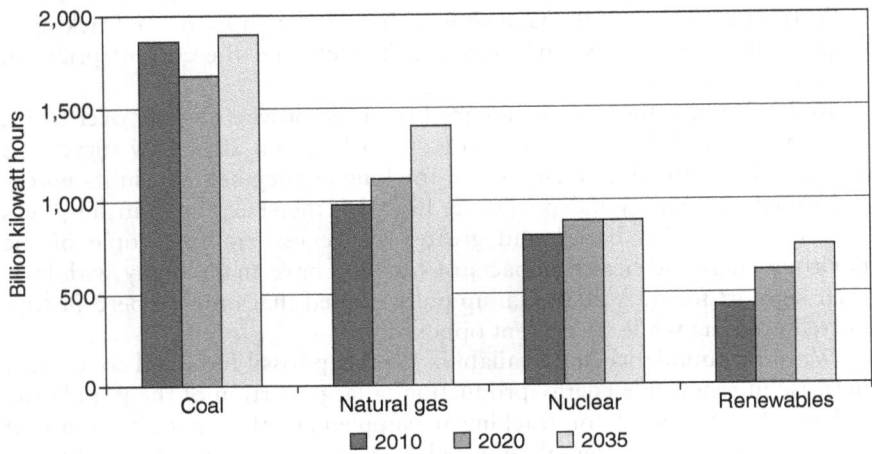

Figure 2.2A Electricity Generation from Different Fuel Sources.
Source: EIA, Annual Energy Outlook, 2012.

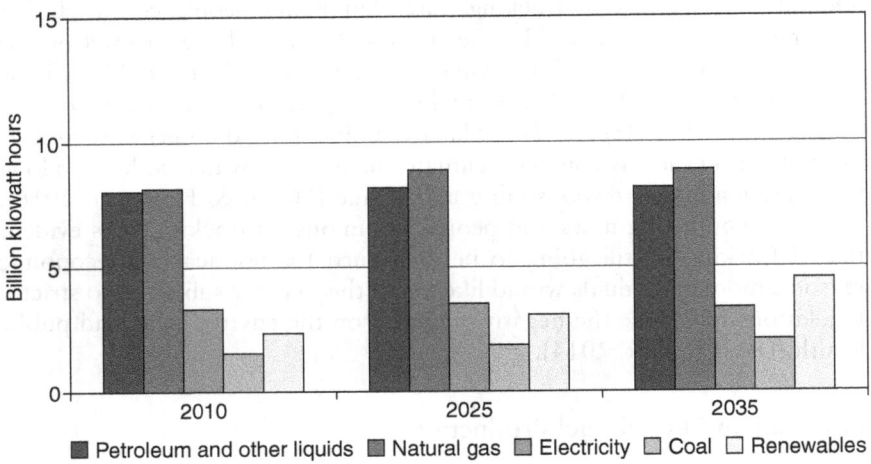

Figure 2.2B Industrial Sector Energy Demand.
Source: EIA, Annual Energy Outlook, 2012.

a transition from coal to natural gas in fueling its generators. As a result, coal now accounts for only 29 percent of electricity production while the share of natural gas has risen to 34 percent (EIA, 2016b). The substitution of coal with natural gas in the utility industry has helped to reduce emissions of carbon dioxide from power plants. The reduced emissions have convinced many people of its potential to mitigate global climate change while others have expressed skepticism.

Additionally, natural gas has posed a threat to the sustainability of nuclear power plants in the nation. The availability of cheap natural gas has made it difficult for some nuclear power plants to remain competitive in the utility market and continue their power supply to the regional electric grid. In Pennsylvania, the owner of the famous Three Mile Island nuclear power plant, Exelon Corp., has announced its plan to close the plant in 2019, if the state fails to come to its financial rescue. Natural gas has dealt a heavy blow to the survival of this plant, a feat which was not even possible by the threats of a meltdown in 1979 (Levy, 2017).

Fracking has also helped to boost crude oil production. Since 2011, the shale formations of Bakken in North Dakota, as well as Permian Basin and Eagle Ford in Texas, have helped to expand crude oil production from 5.6 million barrels per day (bbl/d) to 7.5 million bbl/d in 2013 and 8.7 million bbl/d in 2014. Production increased in 2015, and the trend is expected to continue into 2016. The expansion in domestic crude oil production has led to a 30 percent decline in its import from 2008 to 2015 (EIA, 2016b).

With the increase in crude oil and natural gas production in the United States, the re-entry of Iranian oil into the international market, and reluctance by Oil and Petroleum Exporting Countries (OPEC) to cut down their production, the global inventory of liquefied oil and natural gas has continued to increase while the price has decreased. The excess production of oil has led to a decline in the price of a barrel of oil from over $100 in 2014 to less than $35 at the beginning of 2016. Unexpectedly, the low oil prices have not triggered a consequent increase in its consumption in the global market. Such a phenomenon can be partly attributed to the slow

Table 2.2 Oil and Gas Discoveries in United States Since 2006

Shale	Year Discovered	Location	Type	Estimated Recoverable in Barrels of Oil Equivalent
Marcellus	2006	Onshore	Gas	47 billion
Eagle Ford	2009	Onshore	Oil	23 billion
Wolfe Camp	2010	Onshore	Oil and Gas	8.7 billion
Three Forks	2007	Onshore	Oil	8.4 billion
Utica	2010	Onshore	Gas	7.9 billion

Source: HIS Markit, published in Houston Chronicle on September 8, 2016.

increase in demand for oil and gas in the United States, which is the largest consumer in the world (EIA, 2016b), and the lower economic growth in the emerging markets of Asia and other parts of the world.

As a result of the volatility in the global market, the larger and small oil and gas companies in the United States have shrunk their budgets and expenditures, cut jobs of both unskilled and skilled workers, and reduced rig counts in the oilfields. Additionally, many drilling companies lost their contracts from oil and gas companies and faced economic hardships. This has made them declare bankruptcy while others have consolidated to stay in business. It was amidst such a crisis in the energy sector that oil drillers in Texas broke their 43-year record in crude oil production in 2015 (Eaton, 2016).

History of Fracking in Texas

Drilling for oil and gas has always been a lucrative activity in the state of Texas. According to an Energy Information Administration (2016a) estimate, Texas has the largest reserve of crude oil and natural gas in the nation. The first oil well was drilled in Titusville, Pennsylvania in 1859; drilling activity finally spread to Texas in the 1860s. According to the American Oil and Gas Historical Society (AOGHS, 2015), the first commercial oil well in Texas was drilled in 1866 at Melrose in Nacogdoches County. This was one of the first sites discovered west of the Mississippi river. Explorations led to the discovery of another oil well in 1894 in Corsicana, located just 55 miles south of Dallas in Navarro County. By 1896, there were more wells discovered in Corsicana, and by 1897, it boasted 47 wells, producing 65,975 barrels per day.

The gradual advancements in drilling technology, mainly vertical drilling and a new hydraulic rotary drilling process, enabled and enhanced the commercial production of oil in Texas. With the discovery of more oil wells in Corsicana, the presence of a rig in nearly every house's backyard became a common sight by 1900. The economy of Corsicana experienced a boom, helping transform this regional agricultural shipping center into an oil and industrial center which boasted the highest per capita income in Texas in 1953 (AOGHS, 2016).

In 1901, another major oil field discovery was made in Beaumont, located 90 miles east of Houston in Jefferson County. Here, oil gushed out of Spindletop Hill after an explosion, shooting the liquid hundreds of feet into the air. Nine days after the initial explosion, the oil still gushed out 200 feet into the air, yielding approximately 100,000 barrels a day. Soon six wells were erected on the hill site and Spindletop Hill became the largest oil producing field in the nation. As news spread, many oil drilling companies flocked to the site and they soon changed the landscape of this small lumber town into a major oil production center (City of Beaumont, 2016).

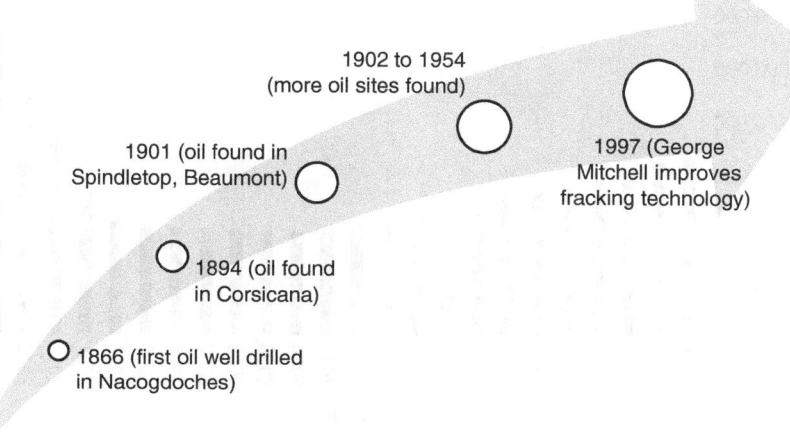

1902 to 1954
(more oil sites found)

1901 (oil found in
Spindletop, Beaumont)

1997 (George
Mitchell improves
fracking technology)

1894 (oil found
in Corsicana)

1866 (first oil well drilled
in Nacogdoches)

Figure 2.3 Timeline in Production of Crude Oil and Natural Gas in Texas.

With the discovery of new oilfields in Texas, oil and gas companies undertook detailed studies of the state's geological formation in search of more potential oil and natural gas reserves. The geological investigations yielded results and provided evidences of more potential oil reserves. It led to explorations from 1902 to 1930, which helped to discover new oilfields in North Central Texas and East Texas (Ramos, 2001). With the discovery of new sites, drilling activities spanned over 65 counties of the state in 1954 and triggered small booms in local economies. With one-third of the nation's oil reserves, Texas' oil production peaked at 3.4 million barrels per day in 1972 and helped to transform the state's agricultural economy into an industrial one.

Crude Oil Production

After reaching the peak in 1972, the state's crude oil production declined to one-third of its peak amount in the later years of the 1980s and the beginning of the 1990s. In 1997, advancements in drilling technology, both vertical and horizontal, coupled with improvements in the fracking process, made possible the commercial production of oil from unconventional sources like shale rock in the state. With the identification of shale plays or formations in the state, the number of oil wells drilled increased along with crude oil production during the period of 2005 to 2014. According to a 2016 EIA estimate, the state has 20 of the top 100 oil

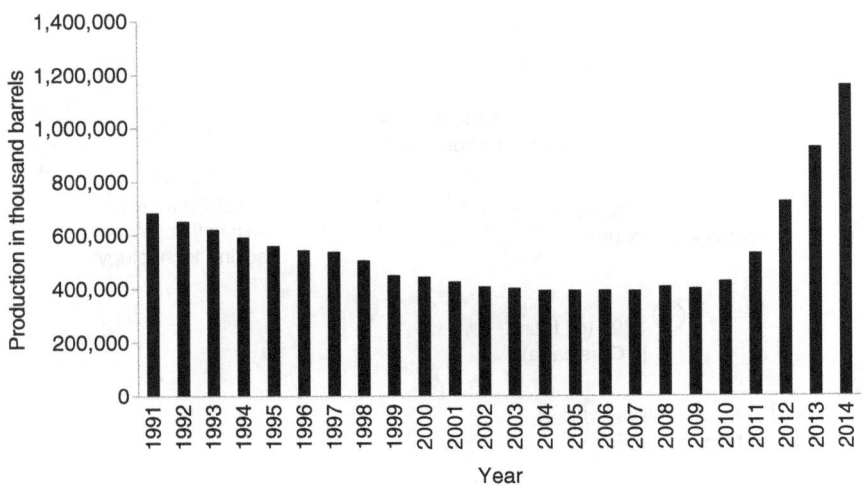

Figure 2.4A Trend in Production of Crude Oil in Texas from 1991–2014.
Source: EIA (2016), Texas Crude Oil Production (thousands of barrels per day).

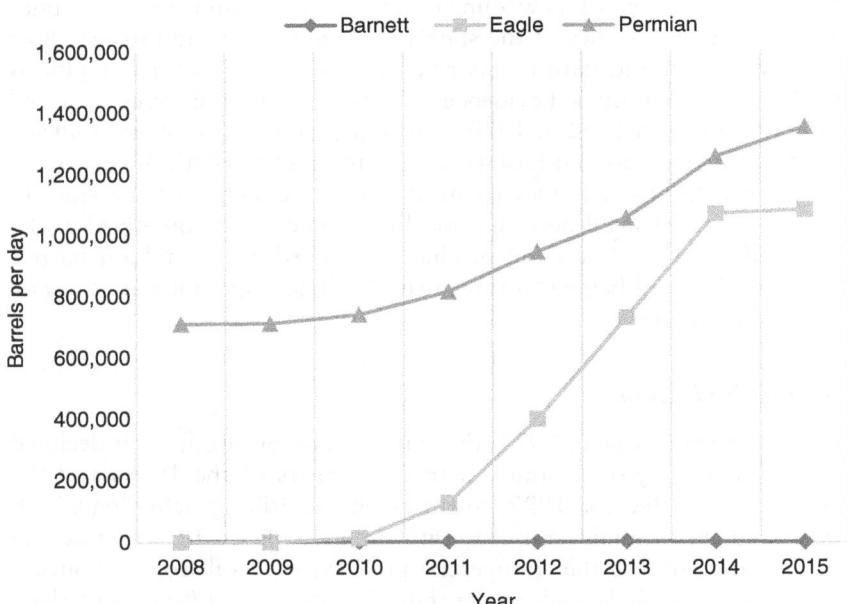

Figure 2.4B Crude Oil Production from Shale in Texas.
Source: Rail Road Commission of Texas (2016), Major Oil and Gas Formations.

producing wells in the nation. These wells have helped the state to maintain its status of being the largest producer of crude oil in the nation, exceeding the nation's offshore crude oil production.

At the beginning of the twenty-first century, crude oil production declined but subsequently gradually increased and reached the level of 3.1 million barrels per day in 2014. In Texas, the most productive oil fields are located in the Permian Basin of West Texas. Even with the decline in oil prices in 2016 to less than $35 a barrel, there has not been a significant decline in crude oil production. Oil companies have been reluctant to cut back their production as 90 percent of their production costs are paid with borrowed money, which makes the oil business both cyclical and volatile in nature. To pay off loans, distribute dividends to shareholders, and prevent bankruptcy, the oil companies need to continue their production despite the glut in the oil market (Tomlinson, 2016).

Natural Gas

Natural gas is another important component in the production of the oil and gas industry. The United States has proven reserves of 3 percent of the world's supply of natural gas and accounts for 22 percent of the nation's energy consumption. About half of these natural gas reserves are located in the states of Texas and Louisiana (EIA, 2016a).

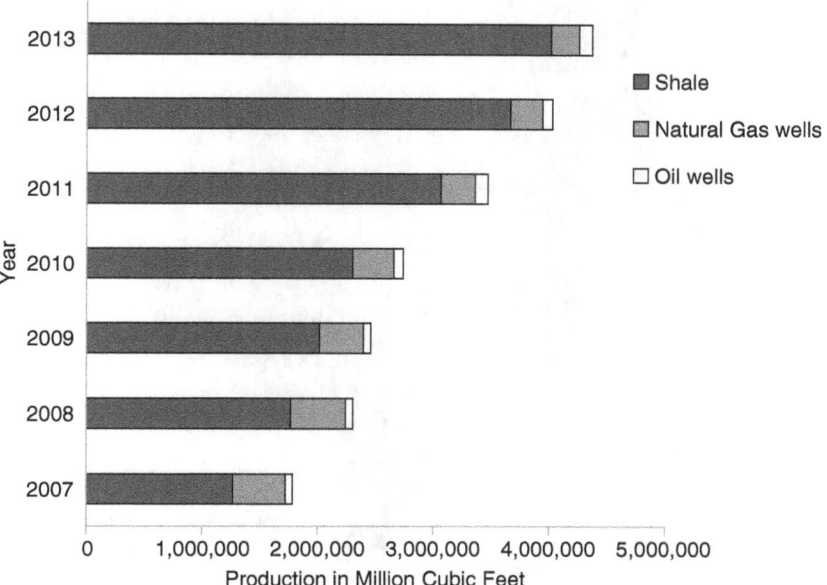

Figure 2.5 Natural Gas Production from Different Sources in Texas.

Source: EIA (2016), Natural Gas.

In Texas, natural gas was initially extracted as a by-product of oil. As a result, its production was dependent on that of crude oil. Later, the discovery of natural gas fields in various parts of the state helped to increase production, which reached a peak of 9.6 trillion cubic feet in 1972 (State-Impact, 2016). In the 1980s and 1990s, production of natural gas declined but continued to remain steady even at a lower level. In 1997, the improvements in fracking technology pioneered by Mitchell Energy and Development Corporation of Houston led to further explorations and the discovery of more natural gas reserves in the shale plays of the state.

Out of the 254 counties in Texas, there are only 27 counties which do not have natural gas wells, as depicted by the gray areas in Figure 2.6. Usually, the shale wells range in depth from five feet to a few thousand feet below ground and produce both natural gas and oil, with the exception of the Haynesville/Bossier Shale which produces only natural gas. The Haynesville formation at the border of Texas and Louisiana accounts for 10 percent of the nation's natural gas reserves and is noted

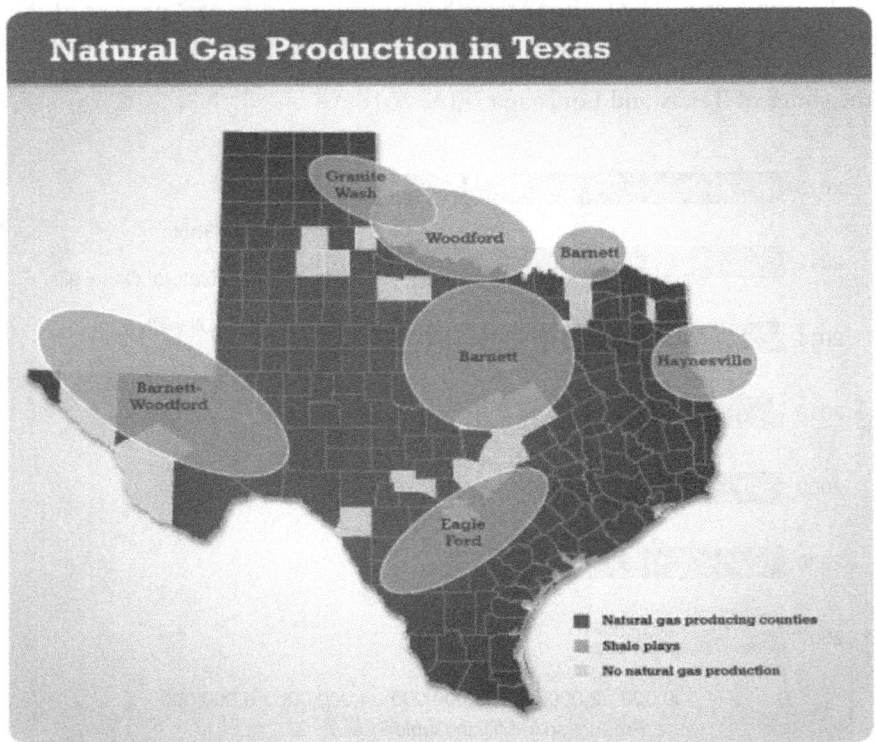

Figure 2.6 Distribution of Shale Gas Reserve in Texas.

Source: www.texasnaturalgasnow.com/natural-gas-in-texas/regional-spotlights.

for the highest production of natural gas. The Barnett Shale formations, covering 5,000 square miles in North Central Texas and stretching across 18 counties, account for 6 percent of the nation's natural gas reserves and are the second highest producers of natural gas in the state (Texas Rail Road Commission, 2016a; Intek, 2011). Other prominent shale formations in the state include the Eagle Ford in East Texas, extending over a length of 400 miles and a width of 50 miles, and the Permian Basin in West Texas, covering an area of 300 miles long and 250 miles wide.

With shale resources emerging as a formidable source of natural gas supply, they have only helped to boost the state's production of natural gas since 2007. When added to the production of natural gas from existing oil and gas wells, the total production of natural gas has steadily increased over the years. In 2014, the state became the largest producer of natural gas, accounting for 29 percent of the nation's production. Though total production of natural gas slightly declined in 2015, the state has been able to maintain its edge in national production (EIA, 2016a).

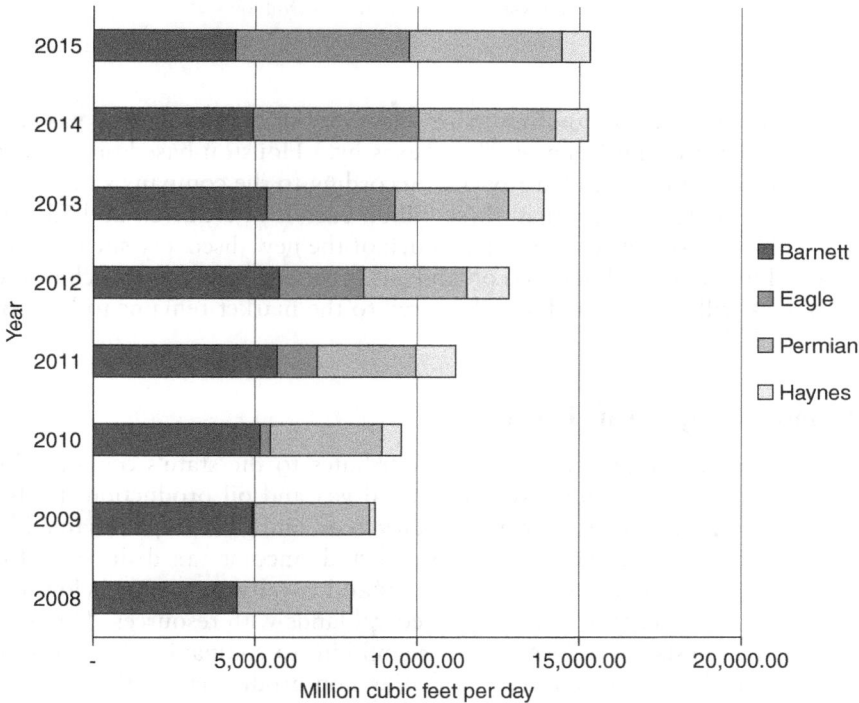

Figure 2.7A Natural Gas Production from Shale in Texas.

Source: Rail Road Commission of Texas (2016), Major Oil and Gas Formations.

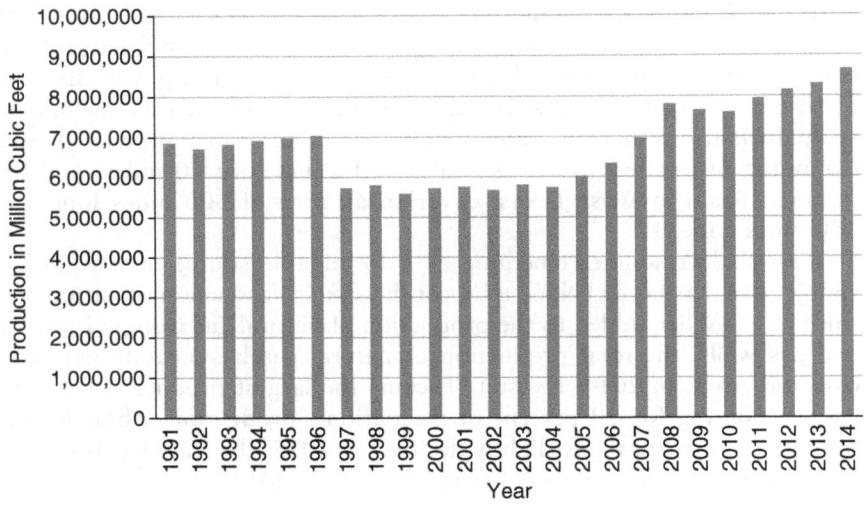

Figure 2.7B Trend in Production of Natural Gas in Texas.
Source: EIA (2016), Texas Natural Gas Gross Withdrawals (Million Cubic Feet).

In September 2016, another major discovery of oil and natural gas was made in the Permian Basin of West Texas by a Houston-based oil and gas exploration company called Apache. According to the company's estimate, the new field holds more than three billion barrels of crude oil and 75 trillion cubic feet of natural gas. Since much of the new discovery site is undeveloped without pipelines, and oil and gas prices are low, how much of the resources will be extracted and delivered to the market remains to be seen (Hunn, 2016).

Economic Impacts in Texas

The oil and gas industry in Texas contributes to the state's economy by creating jobs and paying taxes on natural gas and oil production. It also benefits from the state's offer of subsidies to oil and gas companies in addition to federal ones. Federal incentives include income tax deductions for certain types of oil and gas drilling costs and royalty payments below the market price on lease or purchase of federal lands with resources. The state subsidy manifests in the form of tax exemptions to companies that produce oil and gas from well borings certified as non-producing for the last two years. In 2006, the state and local subsidies given to the oil and gas industry accounted for only 1.5 percent of the total state spending on energy (Texas Comptroller, 2008).

The oil and gas industry has been able to add more jobs to those already existing in extraction, drilling, refining, transportation, and allied industries as a result of discovery of oil and natural gas in the shale plays. During the 2007–2012 period, when the total average annual employment in the United States declined by 2.7 percent, Texas saw an increase in employment in the oil and gas industry by 31.6 percent (Bureau of Labor Statistics, 2014). The upward trend continued with the state adding 19,000 new jobs in oil and gas production in 2013 (EIA, 2013) and also in 2014. Most of these new jobs were concentrated in five counties: Midland, Bexar, Frio, Ector, and Tarrant. Meanwhile, both Harris and Collin County experienced job losses due to the movement of oil and gas companies' headquarters to other parts of the state (Headlight Data, 2015).

The increase in employment figures in the oil and gas industry coincided with the increase in value of the crude oil and gas produced in Texas. During 2007 to 2014, the identification of oil and gas rich shale formations coupled with an increase in the price of a barrel of crude oil and natural gas in the global market led to higher production and an increase in the value of both crude oil and natural gas produced in Texas. From 2015 onwards, the decline in the prices of both crude oil and natural gas have led to a descent in the value of crude oil and natural gas produced in the state.

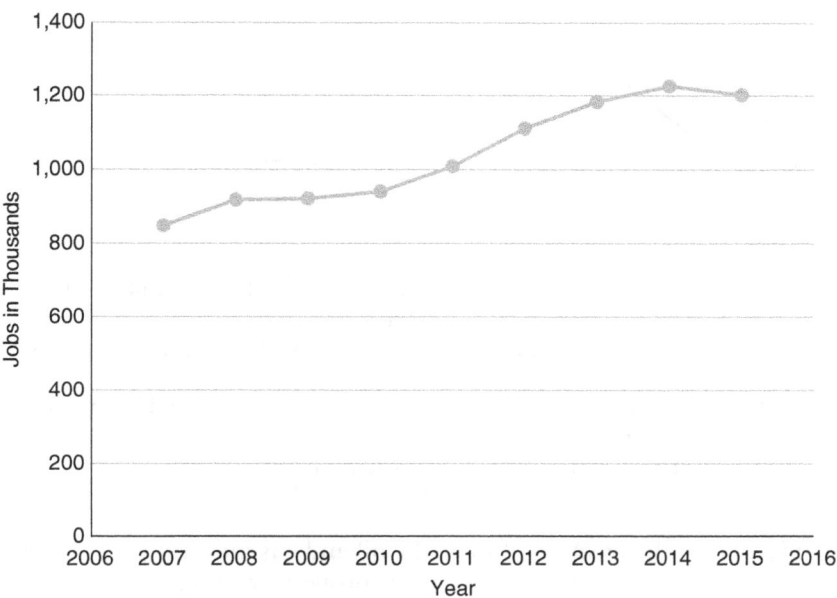

Figure 2.8 Employment Data of Oil and Gas Industry in Texas.

Source: U.S. Federal Reserve Bank, Dallas, Texas.

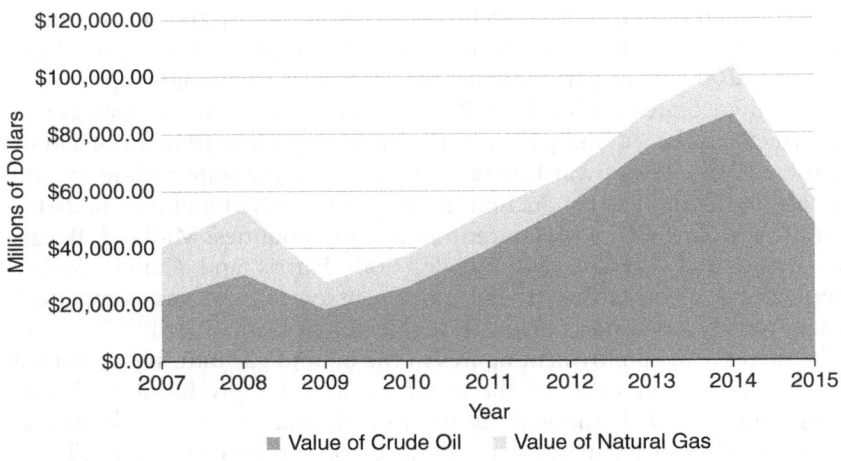

Figure 2.9 Value of Crude Oil and Natural Gas Produced in Texas.

Source: Texas Comptroller's Office, Economic Facts.

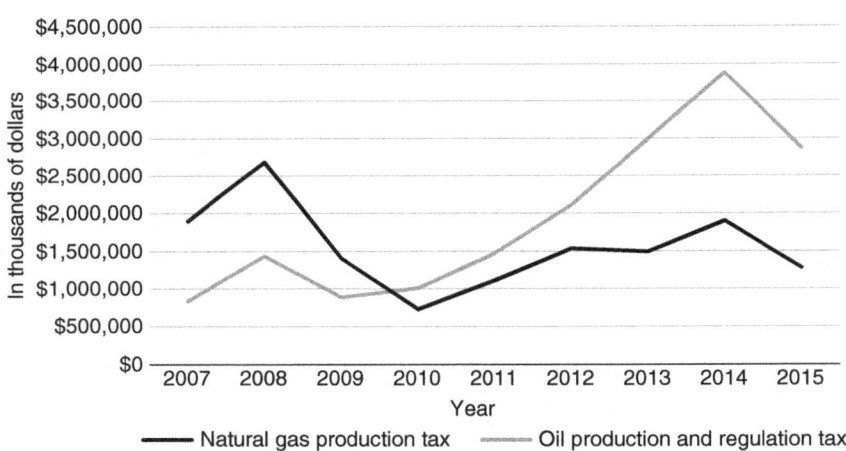

Figure 2.10 Tax Collections from Natural Gas and Oil Productions from Regulations.

Source: Texas Comptroller's Office, 2016, Net State Revenue by Source.

In addition to the creation of jobs, the oil and gas industry pays various types of taxes to the state, ranging from property to severance taxes. A severance tax is contingent on removal of non-renewable resources like crude oil and gas. It includes oil and gas production taxes and oil regulation tax. The state levies a 7.5 percent tax on natural gas production while crude oil production is taxed at the rate of 4.6 percent of its market value. In 2015, the oil regulation tax was repealed by the state's eighty-fourth legislature. The total tax revenue received from natural gas and oil production has varied throughout the years. As shown in Table 1.3, taxes collected from oil and gas production constitute not only a small fraction of state total taxes but also of total state revenue. In 2014, taxes from oil and gas production accounted for 8.17 percent of total state revenue while in 2016, the number dipped to a record low of 3.18 percent with the decline in oil and gas prices, which led to a slowdown in fracking related activities.

Much of the revenue generated by the oil and gas industry is used by the state to fund schools, roads, first responders, and other essential services while state royalty is usually spent on healthcare. The severance taxes imposed on the oil and gas industry make up more than 85 percent of the state's rainy-day fund or the economic stabilization fund (Costello, Green & Graves, 2018). Even the severance fee imposed on the industry for non-compliance with state regulations has helped the state to earn considerable revenue. During the period of 2008 to 2016, the state earned a ten-year average amount of $1,881,481.01 from severance fees. The highest amount of $3,100,525 was collected in 2014. Although the year 2015 marked the beginning of turmoil in the oil and gas market, the state still managed to collect a hefty sum of $2,115,379.50 from severance fees, the third highest amount after the collection of $3,058,625 in 2013 (Texas Railroad Commission, 2016b).

Despite the beginning of a downturn in oil and gas prices in 2015, the industry still paid $13.8 billion in state and local taxes and state royalties, the second highest amount in the state's history. The slide in oil prices has led to a 48 percent decline in state tax collections in the first five months of the fiscal year of 2016 in comparison with the same period the previous year (Texas Comptroller, 2016). Contrary to expectations, the decline in oil prices has not brought about an increase in its consumption by automobiles and other vehicle drivers for many reasons including increased motor vehicle efficiency. This has affected the profit margins of many oil refineries in the state. The refineries buy oil at cheap prices during an oil crash and sell it as gasoline and diesel at a profit. In the absence of an increase in demand, the inventory has expanded leading to serious concerns among investors. Also, the warmer winter in the northeastern United States in 2016 has dealt a blow to the profitability of refining companies.

Adding to the existing woes are economic analysts' expectations of a slow rebound in oil prices, which has led to further devaluation of oil refining companies' stocks by another 2 to 5 percent (Blum, 2016). In 2016, the scenario in oil and gas production remained bleak in the state. With global oil prices hovering around $30 a barrel and OPEC countries still undecided over production cutbacks, fears of bankruptcy remained high among Texas business owners in both energy and non-energy industries. Economists

Table 2.3 Collection of Taxes from Natural Gas and Oil Production and Regulation

Year	Oil Production Tax (in thousands)	Natural Gas Production Tax (in thousands)	Total State Tax Revenue (in thousands)	[1]Total State Revenue (in thousands)	Oil and Gas Production Tax as Percentage of Total State Revenue (in thousands) (%)
2005	681,294 (2.28%)	1,657,086 (5.55%)	29,838,278	43,000,416	5.44
2006	861,659 (2.57%)	2,339,147 (6.97%)	33,544,498	47,694,496	6.71
2007	834,374 (2.26%)	1,895,488 (5.13%)	36,955,630	52,813,239	5.17
2008	1,436,243 (3.47%)	2,684,648 (6.49%)	41,357,929	60,744,698	6.78
2009	883,774 (2.34%)	1,407,739 (3.72%)	37,822,453	53,425,532	4.29
2010	1,008,074 (2.85%)	725,538 (2.05%)	35,368,901	50,500,531	3.43
2011	1,472,111 (3.79%)	1,109,718 (2.86%)	38,856,176	55,840,414	4.62
2012	1,472,111 (4.77%)	1,534,630 (3.48%)	44,079,119	61,739,347	4.87
2013	2,989,542 (6.26%)	1,495,203 (3.13%)	47,781,046	66,512,887	6.74
2014	3,872,279 (7.59%)	1,899,582 (3.73%)	50,992,560	70,676,251	8.17
2015	2,877,020 (5.57%)	1,280,410 (2.48%)	51,683,060	72,727,268	5.72
2016	1,703,903 (3.51%)	578,799 (1.19%)	48,476,226	71,807,036	3.18

Sources: Texas Comptroller of Public Accounts, Sources of Revenue 1972–2016. (a) Natural gas and oil production taxes, p. 147 and p. 151. (b) Sources of Revenue, Appendix A, Tax and Other Revenue Collection, p. 177.

Notes
Numbers in parentheses indicate oil and gas production taxes as percentage of total state tax revenue.
1 Total State Revenue is the sum of total tax revenue and other revenue sources.

predicted that the gloom in the energy sector would lead to more filings of bankruptcy and restructuring of companies in the oil patches of the Permian Basin in the west and the Eagle Ford Shale in the south (Eaton, 2016). Many banks that have loaned money to oil and gas producing companies for drilling may not even be able to recover their loans, creating panic among large and small banks in the region. For example, in the Eagle Ford region, 15 community banks that once experienced a huge swell in their deposits of more than $1.8 billion (a 140 percent increase), from 2008 to 2014, experienced a decline in their deposits. Overall, bank deposits declined from 2014 to 2016 by 3.5 percent, equivalent to a loss of $111 million (Danner, 2016).

In 2017, with the continuation of the OPEC deal that has called for cutbacks in global production, the oil price per barrel has climbed up to $50. The advancements in technology and processes used in oil and gas production have made it possible for energy companies to reap profits even at a lower price level. With the shift of the fracking boom from the Barnett Shale to the Permian Basin of Texas, much of the oil production is concentrated here. The Permian Basin accounts for half of the oil rigs in the nation in 2017. Every addition of a rig helps to create 30 jobs. During a six-month period from November 2016 to April 2017, the oil and gas industry has added 12,000 jobs to its payroll. Despite the gradual job gains in the state's energy sector, total employment in the oil and gas industry still remains a third less than what it was during 2014 (Eaton, 2017).

To provide a steady supply of both skilled and unskilled workers to the energy sector, educational institutions at the community and state levels have partnered with the oil and gas industry to establish programs to meet their labor demands. These programs offer degree and non-degree training programs ranging from engineering and geological sciences to machining, welding, drafting, maintenance technology, automated manufacturing, and data technology (Sweeten, 2016). Even though some demands for jobs in chemical and allied industries still exist, the layoffs in the energy sector have affected the enrollment of students in some of these programs. However, hopes of an increase in demand for skilled and unskilled workers continue to remain high in most of these training programs despite the downward spiral in the oil and gas prices in 2016. Educators and investors continue to anticipate that demand for jobs will pick up by 2018 after a brief hiatus of a year or two. Mostly demand for engineers will increase, followed by that of other skilled and unskilled laborers in various categories of the industry, including those to produce chemical building blocks, maintain existing facilities, and even build new ones based on a new type of demand in the energy industry.

Effects of Sudden Drop in Oil and Gas Prices

In 2016, the recent monetary losses incurred by the oil and gas industry in Texas and other oil and natural gas producing states have been compensated by gains in other non-energy sectors of the economy, both within and outside the state. People have found relief at gas pumps and in payment of their utility bills. Indeed, the lower transportation and utility costs have helped to bolster people's savings. Additionally, the increase in disposable income in the hands of people allows for increased spending on other items. Even non-energy industries have benefited from the drop-in fuel prices through a decline in their costs of production and lower utility payments. Airlines are now paying less for jet fuel and are in a position to pass off some of the savings in the form of lower fares to the public. However, the overall increase in profitability in some of the non-energy industries of the nation has come at the expense of losses worth millions of dollars in the oil and gas industry. Also, the lower oil prices serve as a disincentive to develop alternative sources of energy like wind and solar, essential for sustainable development.

The economic consequences of a precipitous decline in oil and gas prices have been felt mostly by those states whose economy is hinged to oil and gas production. In North Dakota and Alaska, 50 and 80 percent of the state revenue is derived from the oil and gas industry respectively. Other oil and gas producing states with a more diversified economy have been able to fare better despite the losses in jobs and revenue. A good example is the state of Texas. This state, along with its largest city, Houston, has learned valuable lessons from the past to diversify its economy and not rely on a single industry for jobs and revenue. The heavy losses it incurred during the 1980s along with its realization of the cyclical nature and volatility in the oil and gas market has made the state develop plans and implement economic policies geared toward attraction and promotion of other industries providing a variety of products and services. For instance, Baytown, once a booming oil and gas producing city in Texas, is now experiencing growth again after a lull period. Here, the cheap oil and gas is now being used by the refineries and chemical plants as raw material for the production of plastics and other products. The multibillion dollar petrochemical industry that has anchored here has helped to fuel the construction industry. With the creation of new jobs, more homes, apartment complexes, and stores are being built, and the city sales tax revenue at the beginning of 2016 has already risen by 5.6 percent (Collette & Mulvaney, 2016). Thus, by embarking on a strategy of diversification, the state has been able to balance its job losses and reduced revenue from the oil and gas industry with growth in other sectors of the economy.

A closer look at the state's combined revenue from oil and gas production reveals that it is less than 13 percent of the state's total revenue collected during the 2007 to 2015 period. The highest percentage of revenue collected in 2014 accounted for only 12.65 percent of the state's total revenue. Also, in the jobs sector, the state's total job losses with a decline in rig counts and production of oil and gas has only been 1.2 percent, in comparison with 4.3 percent in Wyoming and 2 percent in North Dakota and Oklahoma (Murphy, Plante, & Yucel, 2015).

From 2014 to 2015, Texas lost 24,500 jobs in the oil and gas producing patches of the Permian Basin and Barnett and Eagle Ford Shale regions, but these job losses have been compensated by gains in the state's metropolitan areas of Houston, Dallas, Austin, and San Antonio. During that period, the state's 10.5 percent job losses in mining and extraction have been compensated for by a 4.4 percent job increase in education and health services, 4.6 percent in leisure and hospitality, 2.8 percent in professional and business services, as well as minor gains in construction, government services, information industry, transportation, and other sectors (Texas Workforce Commission, 2015). The Federal Reserve Bank of Dallas predicted in 2016 that despite the losses in the oil and gas industry, Texas will still be able to maintain its job growth, albeit at a slower pace of 1 to 2 percent, and keep its unemployment rate slightly below the national average (Phillips, 2016).

The good news for the oil and gas industry is that with OPEC signing an agreement in November 2016 to cut its production by 4.5 percent, oil prices have made a gradual recovery from $26 a barrel to slightly above $52 a barrel in early 2017. This surge in price has led to an increase in optimism among Texas drillers and the consequent return of 140 rigs to the oil and gas fields in the Permian Basin of the state. Even though the number of rigs and drilling activities in the state's oil and gas fields are nowhere close to that during the peak period of 2014, nevertheless, the surge has helped to add 3,000 oil and gas jobs during the last two months of 2016 (Handy, 2017). The slow recovery that started in 2017 is helping the state to offset the loss of 100,000 jobs that began in the middle of 2014 and continued until OPEC's decision to cut oil production by 1.2 million barrels a day in late 2016.

Conclusion

The small price increase in oil and natural gas that started in late 2016 has made the state's oil and gas companies hopeful of gains from energy production again. The Permian Basin in West Texas has become the largest producer of crude oil in the state while production has declined in the Eagle Ford in South Texas. Since the oil and gas prices slump of 2016, 550 rigs have been added to various drilling sites in the nation out of which the Permian Basin accounts for 240 while Eagle Ford accounts for only 45. With the oil and gas prices still remaining low in 2017, many oil and gas companies have found it harder to make profits and did cut back on their expenditures.

In 2018, with the gradual rise in the price of a barrel of oil to $60 and above in the first quarter of the year, the gloomy outlook in the energy sector has been replaced by signs of recovery. Oil and gas companies have started adding thousands of jobs to the state and national economy while the number of rigs have increased along with oil and gas production. Some of the natural gas produced in the state is now being exported to countries in South America, Europe, Asia, and others from the Gulf ports of Texas. It is also expected that with the transition in fuel type to natural gas in power generation, transportation, and heating and cooling of homes, the demand for natural gas will increase in the near future. Additionally, deregulation of production activities under President Trump's administration has made more land and offshore areas accessible for drilling with fewer environmental restrictions. Despite such a favorable climate for oil and gas drilling, the future of energy production from shale resources faces a threat from an increase in renewable energy production from wind and solar. The latter's market share has risen considerably over the last few years even in the face of tough competition in the energy sector.

References

AOGHS. (2016). First Texas Oil Boom. Retrieved from https://aoghs.org/petroleum-pioneers/texas-oil-boom/.

Blum, J. (2016, February 9). Gasoline supply's rise is alarming investors. *Houston Chronicle*, D2 and D3.

Boudet, H., Clarke, C., Bugden, D., Maibach, E., Roser-Renouf, C., & Leiserowitz, A. (2014). "Fracking" controversy and communication: Using national survey data to understand public perceptions of hydraulic fracturing. *Energy Policy*, 65, 57–67.

Borrick, C. & Clarke, C. (2016). American views on fracking. Center for Local, State and Urban Policy. *Issues in Energy and Environmental Policy*, 28, 1–9.

Bureau of Labor Statistics (2014). Employment changes in the oil and gas industry, by state. Retrieved from www.bls.gov/opub/ted/2014/ted_20140404.htm.

City of Beaumont. (2016). History of Beaumont. Retrieved from www.beaumontcvb.com/about-beaumont/history/.

Collette, E. & Mulvaney, E. (2016, March 13). Booming Baytown props up local economy. *Houston Chronicle*, A1 and A7.

Costello, T. Green, D., & Graves, P. (2018). Fiscal Notes: The Texas Economic Stabilization Fund. Retrieved from https://comptroller.texas.gov/economy/fiscal-notes/2016/september/rainy-day.php.

Danner, P. (2016, February 15). Lower oil prices chip away at Eagle Ford bank deposits. *Houston Chronicle*, B6 and B8.

Davis, C. & Fisk, J. (2014). Energy abundance or environmental worries: analyzing public support for fracking in the United States. *Review of Policy Research*, 13(1), 1–16.

Eaton, C. (2017, June 7). Texas oil jobs stay on the rise, *Houston Chronicle*, B1 and B5.

Eaton, C. (2016, February 7). Oil bounces up again as hopes revive for an OPEC output cut, *Houston Chronicle*, D1 and D2.

EIA (2016a). Texas, state profile and energy estimates. Retrieved from www.eia.gov/state/?sid=TX.

EIA (2016b). Natural gas. Retrieved from www.eia.gov/dnav/ng/ng_sum_lsum_dcu_sTX_m.htm.

EIA. (2013). Texas leads the nation in growth in oil and natural gas production in 2013. Retrieved from www.eia.gov/todayinenergy/detail.cfm?id=18691#.

Gallup. (2016). Opposition to fracking mounts in the U.S. Retrieved from www.gallup.com/poll/190355/opposition-fracking-mounts.aspx?g_source=Politics&g_medium=lead&g_campaign=tiles.

Gertner, J. (2013). George Mitchell. He fracked until it paid off. *New York Times*. Retrieved from www.nytimes.com/news/the-lives-they-lived/2013/12/21/george-mitchell/.

Habib, S. & Hinojosa, M. S. (2016). Representation of Fracking in Mainstream American Newspapers. *Environmental Practice*, 18(2), 83–93.

Handy, R. (2017, January 26). Oil, gas jobs rise again in Texas. *The Houston Chronicle*, B1 and B5.

Headlight Data (2015). Top 10 Counties with best and worst oil, gas and mining economies of 2014: Midland county TX creates most oil and gas jobs, Harris County TX in Houston Loses Most. Retrieved from http://headlightdata.com/top-10-counties-with-best-and-worst-oil-gas-mining-economies-of-2014-midland-county-tx-creates-most-oil-gas-jobs-harris-county-tx-in-houston-loses-most/.

Hilyard, J. (2012). *The Oil & Gas Industry: A Nontechnical Guide*. Tulsa, Oklahoma, PennWell Books.

Hunn, D. (2016, September 8). Apache strikes it big in West Texas: Company believes there's a bounty of oil, natural gas in Permian field. *Houston Chronicle*, A1 & A15.

Hunn, D. (2017, August 2). As the oil patch demands more water, West Texas fights over a scarce. *Houston Chronicle*, A1 and 15. Retrieved from www.houstonchronicle.com/business/article/As-the-oil-patch-demands-more-water-West-Texas-11724100.php.

Intek. (2011). Review of Emerging Resources: U.S. Shale Gas and Shale Oil Plays. A report prepared for EIA. Retrieved from www.eia.gov/analysis/studies/usshalegas/pdf/usshaleplays.pdf Accessed on 2/3/2016.

Levy, M. (2017, May 31). Three mile island could close down. *Houston Chronicle*, B1.

Merrill, T. W. (2013). Four questions about fracking. *Case Western Reserve Law Review*, 63(4), 971–993.

Michaud, K., Carlisle, J. E., & Smith, E. R. (2008). Nimbyism vs. environmentalism in attitudes toward energy development. *Environmental Politics*, 17(1), 20–39.

Murphy, A., Plante, M. & Yucel, M. (2015). Plunging oil prices: A boost for the U.S. economy, a jolt for Texas. Economic Letter of Federal Reserve Bank of Dallas, vol. 10(3), 1–4.

Nicot, J.-P., Hebel, A. K., Ritter, S. M., Walden, S., Baier, R., Galusky, P., Beach, J., Kyle, R., Symank, L., & Breton, C. (2011). Current and Projected Water Use in the Texas Mining and Oil and Gas Industry. Report prepared for Texas Water Development Board. Retrieved from www.twdb.texas.gov/publications/reports/contracted_reports/doc/0904830939_MiningWaterUse.pdf.

Pew Research Center. (2015). Americans, Politics and Science Issues: Climate change and energy issues. Retrieved from www.pewinternet.org/2015/07/01/chapter-2-climate-change-and-energy-issues/.

Pew Research Center. (2015). How Americans view the top energy and environmental issues. Retrieved from www.pewresearch.org/key-data-points/environment-energy-2/.

Pew Research Center. (2012). As gas prices pinch, support for oil and gas production grows. Retrieved from www.people-press.org/2012/03/19/as-gas-prices-pinch-support-for-oil-and-gas-production-grows/.

Phillips, K. (2016). 2016 Texas Economic Outlook: riding the energy roller coaster. Federal Reserve Bank of Dallas Report. Retrieved from www.dallasfed.org/assets/documents/news/releases/nr160112phillips.pdf.

Ramos, Mary G. (2001). Oil and Texas: A cultural history. Texas State Historical Association's Texas Almanac 2000–2001. Retrieved from http://texasalmanac.com/topics/business/oil-and-texas-cultural-history.

Railroad Commission. (2016a). Barnett Shale Information. Retrieved from www.rrc.state.tx.us/oil-gas/major-oil-gas-formations/barnett-shale-information/.

Railroad Commission. (2016b). Permian Basin information. Retrieved from www.rrc.state.tx.us/oil-gas/major-oil-gas-formations/permian-basin/.

StateImpact. (2016c). A look at natural gas production in Texas. Retrieved from https://stateimpact.npr.org/texas/tag/natural-gas-production-in-texas/.

Susquehanna River Basin Commission. (2011). Natural gas development in the Susquehanna River Basin. Retrieved from www.srbc.net/pubinfo/docs/infosheets/Natural%20Gas%20Dev_Fact%20Sheet_FINAL2017.PDF.

Sweeten, V. (2016, February 14). Education cornerstone for careers in energy, *Houston Chronicle*, Fuel Fix.

Texas Railroad Commission. (2016a). Haynesville/Bossier Shale Information. Retrieved from www.rrc.state.tx.us/oil-gas/major-oil-gas-formations/haynesvillebossier-shale/.

Texas Railroad Commission. (2016b). Severance and Seal Orders. Retrieved from www.rrc.state.tx.us/oil-gas/compliance-enforcement/enforcement-activities/severanceseal-orders/.

Texas Comptroller of Public Accounts. (2016). State of Texas. Sources of Revenue: History of State taxes and fees, 1972–2016. Retrieved from www.comptroller.texas.gov/transparency/revenue/docs/96-571.pdf.

Texas Comptroller of Public Accounts. (2008). Government financial subsidies. State Energy Report, Chapter 28, 367–395.

Texas Oil and Gas Association. (2016). Texas Oil and Natural Gas Industry Paid $13.8 Billion in Taxes and Royalties in 2015, Second Most in Texas History. Retrieved from www.txoga.org/texas-oil-and-natural-gas-industry-paid-13-8-billion-in-taxes-and-royalties-in-2015-second-most-in-texas-history/.

Texas Workforce Commission. (2015). Texas Profile – seasonally adjusted. Retrieved from www.tracer2.com/admin/uploadedpublications/1703_TXSadj-Profile.pdf.

Tomlinson, C. (2016, February 9). No good news for oil producers. *Houston Chronicle*, Z9.

3 Environmental Federalism and Federal Regulations on Fracking

Introduction

In the energy producing states of the United States, the federal government has intervened in the exploration and production of oil and natural gas. The concept of federalism has made this intervention possible and it cannot be overlooked in a discussion on federal regulations on fracking. The word "federalism," refers to the sharing of power between the federal and state governments. Disputes in such matters are rampant and continue unabated even today with each level of government threatening to defy each other's orders or practices on grounds of being unconstitutional. Within the concept of federalism, scholars have carved a niche to distinguish a special aspect that deals mainly with the environment. They have labeled this niche "environmental federalism."

Definition of Environmental Federalism

Environmental federalism refers to an allocation system of regulatory authority between the federal and state governments. Under this arrangement, both the levels of government have responsibilities and constraints in their regulatory authority. Historically, states retained the authority to regulate the environment. In those cases of states' failure to address environmental problems that sometimes transcended their boundaries, the federal government has been compelled to intervene into state domains. Such intervention by the federal government caused much dissent among states as they believed they had the constitutional, legal, political, and fiscal capacities to regulate their own environmental woes (Scheberle, 2005). The unsolicited federal interventions resulted in conflicts between states and the federal government that called for judicial intervention. In the settling of such disputes, the Supreme Court has displayed no distinct preference for a particular approach over another. In environmental regulatory matters and depending upon the case, the Supreme Court has ruled in favor of both the federal and the state governments.

Now states continue to make their own decisions on what to regulate and not regulate in specific areas. They simultaneously implement existing federal policies and enforce technology-based standards to protect the air, water, soil, wildlife, and other important aspects of the environment. Only when states fail to implement federal policies does the federal government intervene and use punitive measures to enforce national standards. On the other hand, states retain the right to sue federal regulatory agencies like the EPA if they deem federal standards either to be excessive and not conducive for business or simply absent in protection of public health. For example, the state of Texas celebrated its win in 2012, when the Supreme Court ruled in favor of the state and declared that the EPA had illegally disapproved its pollution permitting program that allowed industrial facilities to bypass burdensome regulations when they reduced air emissions (Satija et al., 2017). On the other hand, in 2006 the state of Massachusetts along with other northeastern states filed a lawsuit in the Supreme Court to protest at the lack of federal initiatives in reduction of greenhouse gas emissions (Heinzerling, 2007).

Sharing of Regulatory Authority between the Federal and State Governments

The arrangement in sharing of responsibility and regulatory authority between the federal and state governments in environmental matters has sparked a debate among scholars over its appropriateness. Scholars have argued both for and against the decentralization of regulatory authority. Those in favor of a centralized approach have validated their stance by arguing that the complex nature of environmental problems and their propensity to transcend jurisdictional boundaries of states justifies federal intervention. Since the federal government has more resources at its disposal along with greater expertise in handling complex environmental problems, it can do a better job than states in addressing the problems (Percival, 1995). Further, decentralization can lead to states' slackening of efforts in enforcement of federal standards and this might trigger a "race to the bottom" in the handling of environmental problems. For example, the EPA under the Obama administration tried to impose stringent regulatory standards on the release of mercury, acid gases, and toxic metals from power plants. The regulation could have prevented 11,000 premature deaths per year and rendered many health benefits to the public but Texas and other states threatened by the regulatory cost of $10 billion that was going to be imposed on the power industry filed a lawsuit against the EPA. The judges in the Supreme Court ruled in favor of the states and called for a revision in federal standards (Satija, 2015). Thus by allowing states to have more say in federal regulatory matters, especially when jobs and corporate investments are at stake in the states' highly competitive business environments, decentralization can sometimes adversely affect the welfare of state residents (Stewart, 1976).

The supporters in favor of decentralization of regulatory authority have refuted those claims. They have pointed out that states are more suited to regulate their environment in an era of unfunded and underfunded mandates. In addition, decentralization comes with significant welfare advantages and there exists no such fear of a race to the bottom arising from interjurisdictional competitions (Tiebout, 1956; Revesz, 1992; Steinzor, 1996; Kraft & Scheberle, 1998). Some scholars have even drawn attention to the positive aspects of decentralization. One of them is that states have served as the birthplace of many novel ideas that have benefited the nation. By acting as the laboratory for innovation and refinement in regulations, the states have actually helped the federal government to adopt those environmental standards and practices that yield positive outcomes and promote public welfare. An example is the federal government's adoption of vehicle emissions and efficiency standards from the state of California's initiatives to curb tailpipe emissions and air pollution (Engel, 2006).

The ensuing debate over the locus of regulatory control has not helped to reach a clear-cut consensus on a single authority that is best suited to address environmental problems or to what extent the federal and state governments should share their regulatory authority. The uncertainty over who is most suited to regulate the environment has helped to produce mismatches between jurisdictional and regulatory authority. As a result, situations that warrant the federal government's intervention lack federal presence, compelling the states to take actions and vice versa. In such circumstances, the distribution of responsibility and regulatory authority between federal and state governments has defied logical reasoning (Adler, 2005).

Overlaps in jurisdictional authority of the federal and state governments also exist. This is evident from the duplicative standards issued by the federal and state governments in pursuit of the same goal. An example is the EPA's issuance of a general standard to attain clean air goals through the national Clean Air Act. This act overlaps with many states' promulgation of higher standards in attainment of their clean air goals, as observed in the states of California, New York, Massachusetts, and others (Esch, 2015). The simultaneous and dual aspects of lack of intervention and existence of overlaps simply add to the dynamism of the concept of environmental federalism.

Those scholars who favor the sharing of regulatory authority have pointed out some of the benefits of such an arrangement. It can lead to optimal regulation through resource sharing and maintain a check and balance on the influence of powerful interest groups. Shared regulatory authority forces interest groups to make a choice at which level of government they should try to exert their influence. Usually interest groups prefer to lobby at the state government level where they can exert a greater level of influence than at the federal level of government (Buzbee, 2005; Engel, 2006).

Due to the complexity of environmental problems and with their numbers rising over the years, some organizations and scholars have made recommendations in the sharing of regulatory authority between the state and federal government. In 1995, the National Academy of Public Administration recommended the formation of new partnerships between federal and state governments because of the many benefits. Such partnerships in an era of devolution of power would help to reduce the EPA's oversight when it is not required. Other recommendations include a multi-tiered arrangement in regulations with governmental participation at local, state, national, and international levels (Esty, 1996), collaborations between various levels of government to attain common objectives through the processes of bargaining, negotiations, and use of advanced management principles, greater reliance on community-based solutions to address environmental problems, reinvention of environmental regulations by the EPA (Kraft & Scheberle, 1998), and placement of limits on federal enforcement ability (Agranoff, 2001). There also exists the suggestion of use of a matching principle in allocation of regulatory authority; that is to match regulatory authority with the geographical scale of the problem (Butler & Macey, 1996). From all these recommendations, what becomes obvious is that they all require the involvement of multiple actors and agencies at various levels of government along with public participation. Incorporating such elements into the concept of environmental federalism can move beyond its static monotonic image to that of a more dynamic one.

Impact of Politics on Sharing of Regulatory Authority

The ensuing conflicts over the locus of control in environmental matters and Republican Party members' demand for decentralization or devolution of power to the states had helped the Reagan administration to advance "New Federalism." This type of federalism called for an expanded role for state governments and the return of regulatory powers to states with little empirical assessments of the impacts of decentralization across policy areas or states' willingness and abilities to assume responsibilities. Also during this time period, the President's Council on Environmental Quality called for the use of cost-benefit analysis to determine the value of environmental regulations and the need for reforms if any, while placing greater reliance on the free market for resource allocation (Lester, 1986).

The drive for devolution of power to states, which initially gained momentum during the Reagan era, lost some of its appeal in the twenty-first century. During times of conflict, some influential Republican Party members have not hesitated to shift their stance from that of a less centralized approach to more centralized control in attainment of specific economic and social goals, as observed during President George W. Bush's administration. In such circumstances, the Republicans not only dismissed their federalism concerns but acted as agents of centralization and encouraged the Democrats to become the advocates of states' authority (Conlan & Dinan, 2007). An example is the federal government's attempt to preempt the power of states in the passage of the Clean Air Interstate Rule and Clean Air Mercury Rule. Through these acts, the EPA sent important messages to the states in 2005. First, states should not establish their own standards to reduce air pollution from particulate matter and ozone. Second, states should use alternative means, like cap and trade programs to limit mercury emissions from power plants and not impose measures that might cause economic hardship to the utility industry (Rabe, 2007).

In addition, the United States' recent withdrawal from the Paris Climate Agreement in 2017 under President Trump's administration has created a new kind of challenge in the sharing of regulatory power. Although the federal government wants to put an end to the sharing of regulatory authority by deregulating or rescinding existing regulations on climate change, not all states are willing to give up their authority in implementing such regulations. These states do realize that governmental inactions would only enhance the risks of damages from ongoing global climate change that they are already experiencing. They also derive support from the majority of the public who believe in climate change and support the government's policy actions. For example, a poll conducted right after the United States' withdrawal from the global climate agreement has shown that 60 percent of Americans believe in climate change. A further breakdown in numbers along party lines has revealed that 83 percent of Democrats, 54 percent of Republicans, and 69 percent of Independents actually agree on evidence of climate change (Borick, Rabe, & Mills, 2017) that warrants action.

The existing complexity in the sharing of regulatory authority gets accentuated when mixed messages are sent by the federal government and it attempts to preempt states' power when it conflicts with its own law. Such tactics have not discouraged the states from pursuing their own regulatory goals. Each state has developed its own framework of regulations to protect the environment and some of these regulations even exceed the threshold standards set by the federal government. For example, when California passed a zero emission vehicle mandate in the 1990s, the automobile industry filed a lawsuit against the state's responsible agency for exceeding the federal standards and won the case, leading to amendments in the mandate.

In regulation of fracking, the federal government's weak stance has only helped the states to further consolidate their position in regulatory matters. With the states taking a lead in the regulation of fracking, the regulatory outcomes have been mixed. The states with strict regulations have either banned or limited fracking much to the disappointment of the oil and gas industry, as observed in the states of New York and Illinois. In other states, either the slack in enforcement of regulations or limited regulations have negatively impacted those individuals who live at close proximity to fracking sites. This has triggered community opposition to fracking and individual lawsuits against oil and gas companies for environmental damages and on grounds of public health concerns, as observed in the states of Pennsylvania and Texas.

Federal Government's Regulations on Fracking

The production and transportation of oil and natural gas from fracking sites to distribution and consumption centers often lead to emissions and leakages of harmful pollutants that pose significant health risks to the public. This has called for government intervention. The states had been responsible for regulating the oil and gas industry's various stages of production from the very beginning. In 1969, the passage of the National Environmental Policy Act (NEPA) paved the way for the federal government's intervention into states' regulatory matters (Merrill, 2012) but this has evoked mixed responses from the states. The states of Texas and Wyoming have welcomed the federal government's intervention through the Halliburton Loophole because it has exempted fracking from the Clean Air and Clean Water Acts. However, states like New York and Florida do not appreciate it.

The proponents of state regulation have pointed out that since the negative externalities arising from fracking are purely local in nature, the responsibility to address them should lie within the scope of the state and local governments. As for the spillover effects of air pollution from fracking sites, the existing Clean Air Act has the provisions to address the problem. With reference to interstate competition and the concern for a race to the bottom, there exist no such fears. Since oil and natural gas bearing shale formations or plays are limited to 33 states, the competition in capital investments is restricted to these states when it comes to oil and natural gas explorations and development. In addition, resource production from shale rocks does not conflict with any national interest. Instead, it helps to meet the nation's energy goal of energy independence, promotes economic prosperity of states and the nation, and reduces greenhouse gas emissions through the substitution of coal with natural gas as the fuel in the utility industry. All these attributes make fracking a good candidate for exemption from federal regulations (Spence, 2012).

The pro-environmental groups consider state regulations on fracking to be inadequate and offering the public unequal protection. They have called for more and stringent federal regulation on fracking to curb the pollution of air, groundwater, and surface water bodies from waste runoff, and to prevent the release of methane from drilling sites. Others in support of federal regulation have further argued that the federal government is better suited to regulate fracking since it has the capability to develop and implement regulations in the best interest of the state and the nation (Garmezy, 2012). Also, with its regulatory headquarters located in the nation's capital at Washington D.C., it does not have to succumb to local pressures from community members and those stakeholders with stakes in oil and gas exploration and production.

Finally, the federal government has justified its decision to intervene by using the Commerce Clause of the U.S. Constitution. This clause gives

Congress the authority to regulate those products that are traded between states and nations. Since there exists an interstate trade in oil and natural gas, it justifies the federal government's intervention and therefore regulation of fracking. To better understand the federal government's regulatory stance on fracking, this chapter has tried to answer the question initially posed in the study, what are the existing federal regulations on fracking?

The Scope of Federal Regulations on Fracking

The federal government has granted the oil and gas industry several exemptions to their fracking operations. These exemptions do not apply to regular oil and gas drilling activities. The industry had secured these exemptions from the federal government on grounds of its unconventionality in extraction of oil and natural gas resources from subsurface shale resources. Some of these exemptions from major federal regulations have been discussed below.

The National Environmental Policy Act (NEPA)

In 1969, President Nixon signed the NEPA. It is not a regulatory statute but a short, simple, and inclusive legislation that aimed at two things – first, environmental protection with the incorporation of environmental values in federal decision making and second, the creation of public awareness through information sharing and public participation. NEPA was also introduced to address the deficiencies that arose from the fragmentation in administrative responsibility in protection of natural resources and the lack of information and expertise in assessment of environment risks in decision making (Dreyfus & Ingram, 1976).

The procedural framework of NEPA requires all federal agencies under Title 1, Section 102 to evaluate the impacts of their actions on the environment and calls for the preparation of analytic environmental impact statements (EIS) and environmental assessments. It offers the public the opportunity to participate and provide inputs into federal decision making by allowing them to attend public meetings and make comments on published drafts of EIS (Environmental Protection Agency, 2016). In recent years, concerns over project delays have encouraged agencies that file the most EIS to streamline their individual NEPA procedures and make use of categorical exclusions.

NEPA even requires independent federal agencies like the Federal Energy Regulatory Commission (FERC) to adhere to its guidelines. FERC oversees the construction of natural gas pipelines and liquefied natural gas terminals and is required by NEPA to conduct a comprehensive review of both upstream and downstream environmental impacts at construction sites. The EPA's Council on Environmental Quality has recommended the assessment of the end use of an authorized project. However, the agency in its rapid approval of projects has declined to conduct comprehensive reviews. Instead, it has narrowed down its focus in assessment of environmental impacts and opted not to assess the impacts of greenhouse emissions on climate change from an approved project (Flyer, 2014). Since FERC's actions have abetted the expansion of fracking infrastructure in urban and rural communities with disregard to the environmental consequences, community residents have launched protests against its pro-corporate decisions. At national and local rallies, sit-ins and blockades at industrial facilities, community residents have demanded their voices be heard in the agency's decision making (Cantarow & Hippouf, 2015).

When it comes to drilling for oil and natural gas, the Energy Policy Act of 2005 (P.L. 109–158 H.R. 6) has also created provisions for streamlining. It has declared that certain oil- and gas-related activities should come under the purview of the departments of interior and agriculture (P.L. 109–158, Section 390) to expedite the process of completing or complying with environmental requirements. The specific activities conducted under the Mineral Leasing Act and for which the aforementioned departments have secured categorical exclusions from NEPA include the following types, as listed under Section 390 of P.L. 109–158 (Yacobucci et al., 2005).

a Individual surface disturbances of less than 5 acres (as long as the
b total surface disturbance on the lease is not greater than 150 acres
c and site-specific analysis in a document prepared pursuant to NEPA
d has been previously completed)
e Drilling an oil or gas well at a location or well pad site at which
f drilling has occurred previously within 5 years prior to the date of
g spudding the well
h Drilling an oil or gas well within a developed field for which an
i approved land use plan or any environmental document prepared
j pursuant to NEPA analyzed such drilling as a reasonably foreseeable
k activity (if that plan or document was approved within the previous
l five years)
m Placement of a pipeline in an approved right-of-way corridor (as
n long as the corridor was approved within the previous five years)
o Maintenance of a minor activity, other than any construction or
p major renovation or a building or facility.

Activities made exempt from NEPA requirements through categorical exclusion preclude public participation. Oil and gas drilling activities on federal lands do not require agencies to address the concerns of nearby landowners. Prior to the passage of the Energy Act of 2005, the burden of proof simply lay with the agency to prove that oil and gas drilling would not cause harm to the public and the environment. Currently, in order to request a NEPA review under extraordinary circumstances (Kosnik, 2007), the public now has to prove that such activities pose a threat to public health and the environment. The exclusion of fracking operations from NEPA review has helped the industry to save both time and money and aided its rapid expansion.

The Safe Drinking Water Act

Federal regulation of drinking water dates back to 1914 when the U.S. Public Health Service declared standards for water quality systems that provided supplies of drinking water to ships, buses, and trains. The initial standards were restricted in scope and applied only to those contaminants that were capable of causing contagious diseases. Subsequently the standards were revised in 1925, 1946, and 1962 to add more contaminants to the list. Even though the standards were not mandated, still some states adopted them to serve as guidelines or as regulations to safeguard water quality.

From the 1960s onwards, a nationwide deterioration in water quality from unrestricted agricultural runoffs, indiscriminate discharges of industrial effluents, and leakages from underground waste disposal sites prompted the federal government to pass the Safe Drinking Water Act (SDWA) in 1974. Under this act, states were required to follow the national safety guidelines and submit data to the EPA on a quarterly basis. To prevent the contamination of public water supplies with harmful injected materials, the SDWA required the EPA to develop minimum standards requirements for underground injection control (UIC). The EPA passed the responsibility to enforce standards and issue permits prior to injection of any fluids underground to the states.

Disputes over states' enforcement of regulations in maintaining drinking water quality have been present for a long time. In 1994, the Legal Environmental Assistance Foundation (LEAF) filed a petition with the EPA to withdraw its approval of Alabama's UIC program. It contended that the state of Alabama had failed to regulate fracking of coalbed methane (CBM) when such a regulation was a requirement under the SDWA to prevent contamination of underground aquifers. The EPA considered LEAF's petition as invalid because fracking did not fall within the statutory or regulatory definition of underground injection (WestLaw, 2010). Unsatisfied with the EPA's response, LEAF filed a petition against the EPA in the eleventh district circuit court in 1997 to make the state of Alabama reform its UIC program. The court concurred with LEAF and asked the EPA to reconsider the issue, but it refused.

The EPA conducted an independent study on injection materials used in extraction of CBM from 2000 to 2004. It revealed little or no threat in contamination of underground sources of potable water. Using its discretionary power, the EPA prevented the states from restricting underground injections used in fracking operations in shale formations, even though its study was restricted to CBM only. The critics expressed their dissatisfaction over the EPA's lackadaisical attitude toward regulation and pointed out the bias in selection of those reports that were in alignment with the agency's views of no conclusive evidence of contamination while neglecting those produced by independent scientific organizations. The EPA was also accused of either removing or changing the levels of risk posed by some chemicals used in fracking, which fell under the purview of SDWA. Regardless of such complaints, the EPA failed to take action (Cupas, 2008).

Later in 2005, with the passage of the Energy Policy Act, Section C of SDWA was completely overturned and fracking for oil and gas received exemption from the regulation. Under the amended SWDA, all fluids used in fracking activities except diesel, a harmful additive, were made exempt from it. Critics have labeled this exemption as the "Halliburton Loophole" because of the leadership role played by then vice president Dick Cheney, who had once served as the chief executive officer of Halliburton, a company well known for its patent on a hydraulic fracturing process in the 1940s and manufacturing of fracking fluids. With a strong bias toward the oil and gas industry, Cheney aided the National Energy Policy Development Committee in securing the exemption. The committee limited participation of environmental groups and their interference in the decision making on exemption. There was also influence exerted by the American Petroleum Institute or API (2016), some of whose recommendations were incorporated word for word into the comprehensive Energy Policy Act (Garmezy, 2012). In finalization of its study, even the EPA was accused of overlooking public comments as well as insights offered by SDWA affiliates such as the Subcommittee on Environment and Hazardous Materials, which has a say in the SDWA (Cupas, 2008).

To address public concerns of possible water contamination at fracking sites and the lurking fear that public pressure on federal government might lead to tightening of existing regulations, the API released a report entitled, "US Oil Shale: Protecting our Environment." In this report, measures undertaken by the oil and gas industry for environmental protection were listed. It also claimed that the oil and gas companies not only comply with federal and state governments' regulations to prevent water contamination but also use advanced technologies like freeze wall to separate underground water from subsurface oil and gas production at fracking sites. Recommendations were made even to adopt alternative strategies in extraction of energy resources from shale deposits that are located much below existing aquifers to minimize the risks of contamination of underground water supply. Despite this, what seems to be lacking from the report and the API's allegations of safety measures undertaken by the oil and gas industry is scientific evidence to refute claims of past studies of the possibility of methane contamination of shallow drinking water aquifers.

A study of private well sites in Pennsylvania and New York, located at close proximity to fracking operations, has revealed that the accumulation of methane over time in enclosed spaces, due to leakages from well casings or migration through an extensive network of fractures in rocks, can pose serious health and fire hazards (Osborn, Vengosh, Warner & Jackson, 2011). Since standards for well casing and enforcement of regulations to protect underground drinking water supply do not necessarily guarantee prevention of leakages (Manuel, 2010), this warrants further investigation and amendment of the SDWA based on valid scientific evidences.

Comprehensive Environmental Response, Compensation, and Liability Act

The Congress passed the Comprehensive Environmental Response, Compensation, and Liability Act (CERCLA) in 1980. Also known as the Superfund Act, this statutory act aims at addressing the problem of potential release and indiscriminate disposal of hazardous chemicals and toxic compounds that pose a danger to public health and the environment through air, water, and soil. It tries to curb such activities by imposing a tax on petroleum and chemical industries along with a penalty for violations of federal regulations.

This regulatory act includes prohibitions and requirements to deal with closed and abandoned toxic waste disposal sites. It tries to seek out responsible parties for environmental damages and hold them liable for cleanups if the government or a private party can establish proof of tangible and intangible losses from the release of harmful substances that have negatively affected the environment and individuals' lives. The act has also established a trust fund to clean up those sites where responsible parties have not been identified. Public involvement in decision making for cleaning up sites is regarded as an essential component of this act.

With the passage of CERCLA, the Congress wanted the public to be involved in decision making and guide the EPA in the cleanup of Superfund sites. The need for community development was further strengthened through the 1986 amendment. The EPA has developed a handbook on community involvement to guide its community involvement effort, and some of its objectives for community involvement include the following (Dreyfus & Ingram, 1976):

- Conduct early, frequent and meaningful community involvement.
- Keep the public well-informed of ongoing and planned activities.
- Encourage and enable the public to get involved.
- Listen carefully to what the public is saying.
- Consider changing planned actions where public comments or concerns are considered by the site team.

- Explain to community members how EPA considers their comments, what the Agency plans to do, and why the decision is made.

Despite the many benefits from the various components of CERCLA, fracking has an exempt status from the liability of CERCLA. Only in the case of release of xylene, toluene, benzene, and other gases that contain elements of petroleum, does the exemption from CERCLA not apply.

Resource Conservation and Recovery Act

The Resource Conservation and Recovery Act (RCRA) aims at the protection of communities from environmental degradation and through resource conservation. It was first enacted in 1965 to address the problems of open burning of waste (Stever, 1986). In 1976, Congress passed the RCRA to address the growing problem of proper and safe disposal of municipal and hazardous industrial waste. Under this act, a compliance mechanism was developed that called for the tracking of waste, recording of movement from origin to disposal site (cradle to grave), and adoption of measures to prevent the entry of hazardous waste into the environment.

The RCRA grants the EPA the authority to issue regulations, provide guidance, and develop policies for the safe disposal of solid and hazardous waste and promote recycling to reduce waste reduction. In implementation of the EPA's policies, both support and cooperation from regional, state, local, and tribal government agencies are essential. To guide agencies, the EPA has developed a comprehensive framework that facilitates waste management and disposal of hazardous and non-hazardous wastes and prevention of contamination at disposal sites. In achieving the objectives of this act, public awareness and participation in decision making are considered essential. The EPA allows the public to participate in decision making in issuance of permits for disposal of hazardous wastes, renewal and modifications of permits, and even in the post closure phase of disposal facilities.

In waste management under the RCRA, waste has been categorized as hazardous and non-hazardous waste. Section C of the RCRA deals with the control of hazardous waste while Section D deals with the management of non-hazardous waste. Section C has a comprehensive regulatory scheme but it allows certain exemptions. The exemptions help waste to be categorized as non-hazardous wastes that are loosely regulated under the RCRA but subject to state or other pertinent federal regulations. The oil and gas industry through intense lobbying influenced Congress to amend Subtitle C with the passing of the Solid Waste Disposal Act in 1980. Under the amendments, various types of oil and gas exploration and production have awaited exemption under Section C, pending a study and regulatory determination by the EPA of the hazardous nature of waste and deficiency in management practices (Environmental Protection Agency, 2002).

The EPA upon completion of its study and under intense lobbying pressure from the oil and gas industry in 1988, submitted its report. The report stated that since the state and federal governments regulated the oil and gas wastes, there existed no need for its regulation under Section C of RCRA. The EPA validated its recommendation on the following grounds: Subtitle C was rigid with little or no considerations for costs and economic impacts and the large volume of oil field waste would exhaust the capacity of Subtitle C waste facilities while permitting delays that would impede the exploration of new oil and gas reserves. The EPA also admitted that based on its study's findings, any mismanagement in disposal of wastes such as land or road spreading of production waste and leakages or seepages from production pits could pose a serious health hazard to the public and the environment (Cox, 2003).

It is because of the exemption that some of the chemicals used in fracturing the shale rocks that have been identified as toxic or carcinogenic when their names were voluntarily submitted by leading oil and gas companies are not subject to control under the RCRA's Section C (46). Even the pits that store flowback water do not require a lining under Section D of the RCRA.

Clean Water Act

The Clean Water Act (CWA) regulates the discharge of harmful pollutants into surface water. Enacted in 1948 under the title of the Federal Water Pollution Control Act, it was reorganized in 1972 and made more elaborate to address public concerns over the growing problems of surface water pollution. The amended act soon came to be known as the Clean Water Act. Under this act, the EPA has the authority to control water pollution through the development and use of wastewater standards for industry and to issue permits for the discharge of pollutants into navigable waters.

Under the federalism framework, even though states set their own standards in alignment with those of the CWA's guidelines, scarcity of resources hinder their enforcement abilities. Such scarcity can be attributed to the small amount of money that is typically set aside in a state's annual budget for environmental protection. To remedy the problem and ensure aggressive enforcement of the act, Congress amended the CWA in 1987. It has granted the EPA the administrative authority to impose two types of criminal penalties, which include "felony" penalties for knowing violations and "misdemeanor" penalties for negligent violations. In this act, the scope of public participation is limited but their expression of concerns for the environment including water pollution and filing of lawsuits against violators does assist the EPA in its enforcement efforts (Rechtschaffen, 2003).

Under the CWA, although the EPA forbids the discharge of oil and other toxic waste into U.S. waters for the risks it poses to public health and the environment, it has granted certain exemptions to the oil and gas industry in its fracking activities. For example, from 1987 to 2005 the EPA has exempted stormwater runoff from oil, gas, and other mining, exploration, and operation sites as long as it is not contaminated through contact with oil field wastes. In 1990, the EPA upon realization that some stormwater runoff may contain sediments from construction sites, considered to treat it as a pollutant if the origin site size exceeded five acres. The site size was revised in 1999 to include sites ranging from one to five acres. Despite the EPA's attempt to control runoff from oil and gas exploration and production sites, the Energy Policy Act of 2005 made amendments in the CWA to exclude sediments as pollutants from both construction and drilling sites. Further, the responsibility to control such runoff has been transferred to state governments, evoking criticisms from environmental groups. They have argued that in the absence of guidance and minimum standards, state efforts can be either lax or stringent depending upon the volume of oil and gas exploration and production activities (Kosnik, 2007).

Clean Air Act

Congress' efforts to control air pollution dates back to 1967 with the passage of the Air Quality Act. This statute enabled the then Secretary of Health, Education, and Welfare to demarcate air quality regions and then assign responsibility to states lying within the regions for the adoption and enforcement of air pollution control standards. Unfortunately, such a regional approach to the control of air pollution failed. When some states were unable to develop an air pollution control program and the public expressed growing concern over air quality, Congress passed the Clean Air Act (CAA) amendment in 1970 (Environmental Protection Agency, 2016h).

The CAA is a detailed federal law aimed at prevention and control of air pollution from stationary and mobile sources. It gives the EPA the authority to establish national ambient air quality standards or minimum air quality standards for criteria pollutants, like carbon monoxide, tropospheric ozone, sulfur dioxide, and total suspended particulates. Other salient features of the 1970 amendment included a goal of 90 percent reduction in emissions from new automobile sources by 1975, use of best available control technology at new sources of air pollution, development of a program to control air toxics, and a strengthened federal enforcement authority. The states were required to develop their own implementation plan to attain and maintain the standards and were also expected to attain air quality standards by 1975 (Rogers, 1990). When many states failed to meet the deadline, the act was again amended in 1977.

In enforcement of the CAA, the public has an important role to play. Under this act, citizens can file lawsuits against corporations or government agencies that disregard emission standards or permit requirements. They can even take legal action against the EPA if it fails to perform or overlooks a non-discretionary action under the act. Despite the many good attributes of the act, it allows unrestricted emissions from scattered oil and gas wells, pipelines, and compressors because they are not aggregated as small sources of emissions. Also, the emissions from these point sources neither fall under the purview of National Emission Standards for Hazardous Air Pollutants nor are subject to control through the Maximum Achievable Control Technology. Even the hydrogen sulfide gas that sometimes leaks from oil and gas well sites and poses a serious hazard to public health is not included in the list of hazardous substances under the act (Kosnik, 2007).

Consequently, the operation of diesel engines at fracking sites and the unrestricted flaring of methane gas at some well pads have led to an increase in emissions of nitrogen oxides and volatile organic compounds into the air (Carlton, Little, Moeller, Odoyo, Shepson, 2014). This is evident in the Marcellus Shale of Pennsylvania and the Bakken Shale of the North and South Dakota and Montana (Prenni et al., 2016). The photochemistry between these two gases has been responsible for an increase in ozone concentration during the winter months in the oil and gas basins of the western U.S. (Edwards et al., 2014) and in the Barnett Shale region of North Texas (Ahmadi & John, 2015).

Emergency Planning and Community Right to Know Act

Congress passed the Emergency Planning and Community Right to Know Act (EPCRA) in 1986. It aims to protect people in their respective communities from the accidental release of toxic chemicals and hazardous gases. Two past incidents had triggered the passage of this act – the accidental release of methylene chloride and aldicarb oxime from a Union Carbide facility at Institute, West Virginia in 1985 and the 1984 Bhopal disaster in India that killed thousands of people from the leakage of methyl isocyanate gas.

The EPCRA is slightly different from other acts in that it does not use a command and control approach. Instead it relies upon the cooperation of companies in reporting the names of hazardous substances they handle, store, and transport. To prevent emergencies, the EPCRA calls for emergency planning at local and state levels, emergency emissions notifications, reporting on storage and transportation of threshold quantities of hazardous chemicals, and submission of an annual report on the release of chemicals over the threshold level that are listed as toxic.

In implementation of the EPCRA, the EPA assumes the role of an intermediary agent between industry and the public in communication of reliable information in a timely manner to prevent emergencies and reduce potential hazards in communities. Upon collection of information, the EPA shares it with the public by posting it online. Its web-based tool called My Right To Know enables the public to access the information via a computer or their mobile smartphone (Environmental Protection Agency,

2016i). However, the act's Section 322 allows a company to obtain an exempt status under its trade secrets clause, which the oil and gas industry has obtained by filing substantiation forms. The exempt status has enabled the oil and gas companies to keep the names of certain proprietary chemicals used in fracking a secret and not disclose information on the types and amount of waste generated at operation sites.

Fracking's exemption from EPCRA reporting requirements does not bode well with community members. Even though there exists an industry website called FracFocus where oil companies can report the names of chemicals used in fracking operations, such reporting is done on a voluntary basis. There still exists the likelihood of companies withholding the names of proprietary chemicals and FracFocus has failed to fill in the void created by the Toxic Release Inventory's (TRI's) exemption rule (McFeeley, 2012). As a result, individuals living in close proximity to fracking sites have no way of finding out the risks associated with certain chemicals used in fracking or from the waste generated by it. In the absence of such

Table 3.1 Federal Regulations and Exemptions on Fracking

Federal Regulations	Exemptions
National Environmental Policy Act (NEPA)	Drilling of oil or gas well at a location where drilling has occurred within 5 years.
Safe Drinking Water Act (SDWA)	Oil and gas companies can inject any fluids other than diesel in fracking related operations (Halliburton Loophole).
Comprehensive Environmental Response, Compensation and Liability Act (CERCLA)	Oil and gas industry is exempt from the act's requirements unless xylene, toluene, benzene, and other gases are involved.
Resource Conservation and Recovery Act (RCRA)	Regulation of oil and gas industry's waste is not necessary under Title C of this act.
Clean Water Act (CWA)	Permit not required for storm water discharges from oil and gas exploration sites and also from construction sites under the 2005 Energy Policy Act.
Clean Air Act (CAA)	Emissions from scattered oil and gas wells are not considered as a 'major source' of air pollution.
Emergency Planning and Community Right to Know Act (EPCRA)	Oil and gas companies do not need to disclose to community members the names of proprietary chemicals used in fracking.
Endangered Species Act (ESA)	None. Adoption of voluntary conservation measures for fracking in critical habitat areas.

information, the public remains at a disadvantage. They do not know what precautions to take to protect themselves and the environment in case of an accident or spill and how to make the oil and gas companies socially responsible for their actions. Even state regulations and exemptions fail to address the issue.

In response to the public and environmental groups' demand for disclosure of chemicals under the Toxic Substances Control Act, the federal government under the Obama administration made feeble attempts at regulation of fracking on federal and tribal lands. In March 2015, the Interior Department's Bureau of Land Management (BLM) announced that all oil and gas companies leasing federal and tribal lands for fracking would have to disclose the names of chemicals within 30 days of starting their operations and follow stringent standards in well casing. As soon as the federal order was announced, the U.S. District Court for Wyoming issued an injunction to block it. Meanwhile the Independent Petroleum Association of America and Western Energy Alliance sued the BLM on grounds of duplicity in standards and therefore lack of necessity. In 2017, the Trump administration rescinded the Obama era rule on the same grounds of duplicity in standards and a high cost of compliance of $45 million per year for the oil and gas industry (Davenport, 2015).

Endangered Species Preservation Act

In 1966, Congress passed the Endangered Species Preservation Act to list those native animal species that were deemed endangered and so protect them from extinction. The departments of interior, agriculture, and defense were assigned the responsibilities to list the protected species and preserve their habitats, while the U.S. Fish and Wildlife Services was authorized to acquire land for purposes of preservation. In 1969, the act was amended to enhance its scope of protection, which led to the inclusion of even those species that faced the threat of worldwide extinction.

A few years later in 1973, Congress passed the Endangered Species Act (ESA). The act made clear its objectives through its definitions of "endangered" and "threatened species," made all plants and invertebrates eligible for protection, clarified the "take" prohibitions applicable to endangered species, and required all federal agencies including the EPA to conserve protected species and refrain from authorizing those activities that might harm listed species or destroy or modify critical habitats. To attain its goals, the ESA provides funding for land acquisition for foreign species and makes matching funds available to states that have cooperative agreements. The ESA is administered by the services comprising of the U.S. Fish and Wildlife Service and the National Oceanic and Atmospheric Administration's National Marine Fisheries Service (Environmental Protection Agency, 2017). The ESA was amended in 1978, 1982, 1988, and 2004, but its original framework was left intact (U.S. Fish and Wildlife Service, 2017).

Unlike other federal regulations, the ESA does not provide any exemption to the oil and gas industry for fracking on land. This unconventional drilling procedure, which requires a larger land area for its well pad sites than the conventional drilling process, does sometimes pose a threat to plant and animal species that have either been listed or are proposed to be listed as protected or endangered. Permission to continue fracking in such circumstances requires the validation of its importance in the state and the national economy, as well as a compromise with the Fish and Wildlife Service. This is evident in Texas and New Mexico, where both states have adopted strategies to keep the dunes sagebrush lizard in the Permian Basin off the ESA listing as protected.

Since the habitat of the lizard is limited to these two states, they have shown a commitment to adopt measures to cut back on the damaging effects of fracking on their habitat and also mitigate the loss of habitat. In such an endeavor, the states have formed coalitions with private landowners, royalty owners, agriculture and oil and gas industries, academia, and representatives from state and federal agencies. Mitigation goals have been developed along with an adaptive management plan that is based on feedback and updated information from the monitoring and review of existing practices in conservation efforts.

Although some environmental organization groups do not like the idea of a compromise between a federal agency and states on the development of a conservation plan to prevent the listing of plant and animal species under the ESA, such voluntary conservation measures are being used by the industry to continue fracking in critical habitat areas (Eckstein & Snyder, 2013). From such a scenario, it has become apparent that even in the absence of an exemption, a deal with the enforcement agency can help to overcome some of the challenges posed by the ESA in fracking.

Conclusion

In the sharing of regulatory power between the federal and state governments, exemptions and loopholes in existing federal regulatory exemptions have only weakened the federal government's regulatory power. In the case of exemptions, since many of them were approved by the federal government at a time when fracking was still at a fledgling stage and some of its environmental impacts were unknown, there exists the need to review them in the current context of externalities arising from fracking practices. Upon investigation, even if scientific evidence warrants the removal of some of these exemptions, it would be difficult to do so. There is intense lobbying by the oil and gas industry which resists changes in regulation, let alone the introduction of new ones.

The lack of bipartisan support also stands in the way of changes in regulations on fracking. Throughout the history of regulations on the oil and gas industry, the existing lack of consensus between Democratic and Republican Party members in enforcement of federal regulations has only benefited the industry, leaving the public at a disadvantage from the negative externalities associated with oil and gas exploration and production activities. Further in a mode of deregulation under the Trump administration, the future of federal regulations on fracking remains uncertain. This has added to the dilemma of those states committed to environmental protection both in dealing with deregulation and accepting greater responsibility in environmental and public protection with less intervention from the federal government.

References

Adler, J. H. (2005). Jurisdictional mismatch in environmental federalism. *NYU Environmental Law Journal*, *14*, 130.

Agranoff, R. (2001). Managing within the matrix: Do collaborative intergovernmental relations exist? *Publius: The Journal of Federalism*, *31*(2), 31–56.

Ahmadi, M. & John, K. (2015). Statistical evaluation of the impact of shale gas activities on ozone pollution in North Texas. *Science of The Total Environment*, *536*, 457–467.

American Petroleum Institute. (2016). U.S. oil shale: Protecting our environment. A report from the API. Retrieved from www.api.org/Oil-and-Natural-Gas-Overview/Exploration-and-Production/Oil-Shale/Protecting-Environment.

Borick, C., Rabe, C., & Mills, S. (2017, June 12). Trump's global warming views remain elusive, but not those of Americans. A Brookings Report. Retrieved from www.brookings.edu/blog/fixgov/2017/06/12/trumps-global-warming-views-remain-elusive-but-not-those-of-Americans/?utm_campaign=Governance%20Studies&utm_source=hs_email&utm_medium=email&utm_content=53124520.

Brady, J. & Crannell, J. (2012). Hydraulic fracturing regulation in the United States: The laissez-faire approach of the federal government and varying state regulations. *Vermont Journal of Environmental Law, vol. 14*, 40–70.

Butler, H. N. & Macey, J. R. (1996). Externalities and the matching principle: the case for reallocating environmental regulatory authority. *Yale Law & Policy Review*, *14*(2), 23–66.

Buzbee, W. W. (2005). Contextual environmental federalism. *NYU Environmental Law Journal*, *14*, 108.

Cantarow, E. & Hippouf, D. (2015). The federal agency behind the gross expansion of pipelines. October 17, 2015. Retrieved from www.truth-out.org/news/item/33239-the-federal-agency-behind-the-gross-expansion-of-fracking-pipelines.

Carlton, A., Little, E., Moeller, M., Odoyo, S., & Shepson, P. (2014). The data gap: Can a lack of monitors obscure loss of Clean Air Act benefits in Fracking Areas? *Environmental Science and Technology*, Viewpoint.

Conlan, T. & Dinan, J. (2007). Federalism, the Bush Administration, and the Transformation of American Conservatism. *Publius: The Journal of Federalism*, vol. *37*(3), 279–303.

Cox, J. R. (2003). Revisiting RCRA's Oilfield Waste Exemption as to Certain Hazardous Oilfield Exploration and Production Wastes. *Villanova Environmental Law Journal*, *14*, 1, 1–38.

Cupas, A. C. (2008). Not-So-Safe Drinking Water Act: Why We Must Regulate Hydraulic Fracturing at the Federal Level, The. *William and Mary Environmental Law and Policy Review*, *33*(2), 605–632.

Davenport, C. (2015, September 30). Judge blocks Obama Administration rules on fracking. *New York Times*. Retrieved from www.nytimes.com/2015/10/01/us/politics/judge-blocks-obama-administration-rules-on-fracking.html?mcubz=0.

Dreyfus, D. & Ingram, H. (1976). National Environmental Policy Act: A view of intent and practice, *The Neural Resources Journal*, 16, 243.

Eckstein, G. & Snyder, J. (2013). Endangered Species in the Oil Patch: Challenges and Opportunities for the Oil and Gas Industry, *Texas A & M Law Review*, 1, 379–409.

Edwards, P. M., Brown, S. S., Roberts, J. M., Ahmadov, R., Banta, R. M., Dubé, W. P., & Helmig, D. (2014). High winter ozone pollution from carbonyl photolysis in an oil and gas basin. *Nature*, 514(7522), 351–354.

Engel, K. H. (2006). Harnessing the benefits of dynamic federalism in environmental law. *Emory Law Journal*, 56, 159–188.

Environmental Protection Agency. (2017). About the Endangered Species Protection Program. Retrieved from www.epa.gov/endangered-species/about-endangered-species-protection-program.

Environmental Protection Agency. (2016a). What does NEPA require? Retrieved from www.epa.gov/nepa/what-national-environmental-policy-act#NEPArequirements.

Environmental Protection Agency. (2016b). EPA Compliance with the National Environmental Policy Act. Retrieved from www.epa.gov/nepa/epa-compliance-national-environmental-policy-act.

Environmental Protection Agency. (2016c). Environmental justice considerations in the National Environmental Policy Act process. Retrieved from: www.epa.gov/nepa/environmental-justice-considerations-national-environmental-policy-act-process. Accessed on 4/8/2016.

Environmental Protection Agency. (2016d). Superfund Community Involvement Handbook.

Environmental Protection Agency. (2016e). History of the Resource Conservation and Recovery Act (RCRA). Retrieved from www.epa.gov/rcra/history-resource-conservation-and-recovery-act-rcra.

Environmental Protection Agency. (2016f). Resource Conservation and Recovery Act (RCRA) overview. Retrieved from www.epa.gov/rcra/resource-conservation-and-recovery-act-rcra-overview. Accessed on 4/18/2016.

Environmental Protection Agency. (2016g). History of Clean Water Act. Retrieved from www.epa.gov/laws-regulations/history-clean-water-act.

Environmental Protection Agency. (2016h). Summary of the Clean Air Act. Retrieved from www.epa.gov/laws-regulations/summary-clean-air-act.

Environmental Protection Agency. (2016i). What is EPCRA? Retrieved from www.epa.gov/epcra/what-epcra.

Environmental Protection Agency. (2016j). My right to know. Retrieved from www.epa.gov/toxics-release-inventory-tri-program/my-right-know-application.

Environmental Protection Agency. (2002). Exemption of oil and gas exploration and production wastes from federal hazardous waste regulation. Retrieved from www3.epa.gov/epawaste/nonhaz/industrial/special/oil/oil-gas.pdf.

Environmental Protection Agency. (1999). Twenty five years of the Safe Drinking Water Act: History and trends. A Report from the Office of Water, 1–56. Retrieved from http://permanent.access.gpo.gov/websites/epagov/www.epa.gov/safewater/consumer/trendrpt.pdf.

Environmental Protection Agency. (1996). RCRA expanded public participation rule. Retrieved from www.epa.gov/sites/production/files/2015-08/documents/brochpdf.pdf.

Esch, M. (2015, June 29). New York state formalizes ban on fracking, ending 7-yr review. Associated Press. Retrieved from http://infoweb.newsbank.com/resources/doc/nb/news/15646D6174972E38?p=AWNB.

Esty, D. C. (1996). Revitalizing environmental federalism. *Michigan Law Review*, 95(3), 570–653.

Flyer, A. (2014). FERC Compliance Under NEPA: FERC's Obligation to Fully Evaluate Upstream and Downstream Environmental Impacts Associated with Siting Natural Gas Pipelines and Liquefied Natural Gas Terminals. *Georgetown International Environmental Law Review*, 27, 301.

Garmezy, A. (2012). Balancing hydraulic fracturing's environmental and economic impacts: the need for a comprehensive federal baseline and the provision of local rights. *Duke Environmental Law and Policy Forum*, 23, 405–439.

Heinzerling, L. (2007). Massachusetts v. EPA. *Journal of Environmental Law and Litigation*, 22, 301.

Kosnik, R. (2007). The oil and gas industry's exclusions and exemptions to major environmental statutes. Oil and Gas Accountability Project Report. Retrieved from www.ogap.org.

Kraft, M. E. & Scheberle, D. (1998). Environmental federalism at decade's end: New approaches and strategies. *Publius: The Journal of Federalism*, 28(1), 131–146.

Lester, J. (1986). New federalism and environmental policy. *Publius: The Journal of Federalism*, 16(1), 149–166.

Manuel, J. (2010). EPA tackles fracking. Environmental Health Perspectives, vol. 118(5), A199.

McFeeley, M. (2012). State hydraulic fracturing disclosure rules and enforcement: A comparison. A NRDC Issue Brief. Retrieved from www.nrdc.org/sites/default/files/Fracking-Disclosure-IB.pdf.

Merrill, T. W. (2012). Four questions about fracking. *Case Western Reserve Law Review*, 63(4), 971–993.

National Academy of Public Administration. (1995). Setting priorities, getting results: a new direction for the Environmental Protection Agency. Report to Congress (Washington D.C., National Academy of Public Administration, 1995), 2.

Osborn, S. G., Vengosh, A., Warner, N. R., & Jackson, R. B. (2011). Methane contamination of drinking water accompanying gas-well drilling and hydraulic fracturing. *Proceedings of the National Academy of Sciences*, 108 (20), 8172–8176.

Percival, R. V. (1995). Environmental federalism: historical roots and contemporary models. *Maryland Law Review*, 54, 1141.

Prenni, A. J., Day, D. E., Evanoski-Cole, A. R., Sive, B. C., Hecobian, A., Zhou, Y., & Schurman, M. I. (2016). Oil and gas impacts on air quality in federal lands in the Bakken region: an overview of the Bakken Air Quality Study and first results. *Atmospheric Chemistry and Physics*, 16(3), 1401–1416.

Purifoy, D. M. (2013). EPCRA: A Retrospective on the Environmental Right-to-Know Act. *Yale Journal of Health Policy Law & Ethics*, 13, 375–416.

Rabe, B. (2007). Environmental policy and the Bush era: The collision between the administrative presidency and state experimentation. *Publius: The Journal of Federalism*, 37(3), 413–431.

Rechtschaffen, C. (2003). Enforcing the Clean Water Act in the twenty-first century: Harnessing the power of the public spotlight. *Alabama Law Review*, 55, 775–814.

Revesz, R. L. (1992). Rehabilitating Interstate Competition: Rethinking the Race-to-the-Bottom Rationale for Federal Environmental Regulation. *New York University Law Review*, 67, 1210–1254.

Rogers, P. (1990). EPA history: The Clean Air Act of 1970. Retrieved from www.epa.gov/aboutepa/epa-history-clean-air-act-1970.

Satija, N., Carbonell, L & McCrimmon, R. (2017, January 17). Texas vs. the Feds – A look at the lawsuits. *Texas Tribune*. Retrieved from www.texastribune.org/2017/01/17/texas-federal-government-lawsuits/.

Satija, N. (2015, June 29). Supreme court thwarts EPA mercury rules in a victory for Texas. *Texas Tribune*. Retrieved from www.texastribune.org/2015/06/29/supreme-court-mercury-ruling/.

Scheberle, D. (2005). The evolving matrix of environmental federalism and inter-governmental relationships. *Publius: The Journal of Federalism*, 35(1), 69–86.

Spence, D. and Adelman, D. (2016, February 12). Texas officials claim victory over Clean Power Plan — at expense of every Texan. *Dallas Morning News*. Retrieved from www.dallasnews.com/opinion/latest-columns/20160212-david-spence-and-david-adelman-texas-officials-claim-victory-over-clean-power-plan-at-expense-of-every-texan.ece.

Spence, D. (2012). Energy management brief: Is it time for federal regulation of shale gas production? *Energy Management and Information Center*. University of Texas, Austin.

Stever, D. W. (1986). *Law of Chemical Regulation and Hazardous Waste*. C. Boardman Company.

Stewart, R. B. (1976). Development of Administrative and Quasi-Constitutional Law in Judicial Review of Environmental Decision making: Lessons from the Clean Air Act, The. *Iowa Law Review*, 62, 713.

Steinzor, R. I. (1996). Unfunded Environmental Mandates and the New (New) Federalism: Devolution, Revolution, or Reform. *Minnesota. Law. Review*, 81, 97–227.

Tiebout, C. M. (1956). A pure theory of local expenditures. *Journal of Political Economy*, 64(5), 416–424.

U.S. Fish and Wildlife Service. (2017). A history of the Endangered Species Act of 1973. Retrieved from www.fws.gov/endangered/esa-library/pdf/history_ESA.pdf.

WestLaw (2010). LEAF v. EPA, 118 F.3d 1467.

Yacobucci, B., Coplena, C., McCarthy, J., Powers, K., Sisssine, F., & Tiemann, M. (2005). Key Environmental Issues in the Energy Policy Act of 2005 (P.L. 109–58 H.R. 6). Congressional Research Report.

4 Factors Affecting States' Decisions to Frack and State of Texas' Regulations on Fracking

Introduction

It is a challenging task for an elected politician to introduce or amend existing regulations based on constituents' requests, saliency of an environmental issue, or any other problem. Also, for any policy change to take place, the "window of opportunity" (Kingdom, 1994) has to open up and decision makers need to act quickly to avail the opportunity of either introducing a new policy or amending an existing one. Ideally, whatever maybe the case, a balance needs to be sought between the economic and social interests of the people. Since economic and social interests are both dynamic in nature, the act of balancing is a daunting task when it is tainted with politics. This can be observed in the clash between the logging industry demanding greater access to forests and environmentalists' wanting to protect the habitat of the spotted owl in the west coast (Barringer, 2007). As a result, a debate on jobs versus environment was initiated and seems to resonate in the case of fracking.

Factors Influencing Regulations

In the economy, whenever a steady stream of revenue is generated by a lucrative business activity and there is strong lobbying by the industry, it becomes difficult to regulate the negative externalities. Often with economic interests taking precedence over environmental protection, there exists the possibility of existing regulations being compromised with approvals of exemptions and management plans showing concurrent collaborations with various stakeholders to curtail environmental pollution. When it comes to the formulation of new regulations, the prevailing political climate and the nation and state's commitment to energy independence and environmental protection determine to some extent the stringency of regulations.

It has been observed that when existing regulations fail to provide adequate protection to people, grass-root organizations and national environmental groups enter the scene and play an important role. They serve as watchdogs over industries that pollute and bring the incidents of environmental abuse to the public and media's attention. Sometimes, such actions are followed by lawsuits to correct the wrongdoing. The long drawn and costly legal battles that ensue usually send messages of warning to polluters, irrespective of who wins the case. Despite such efforts, it should also be kept in mind that some environmental organizations do not hesitate to receive funding from polluting companies whose actions they condemn. For example, the Sierra Club received $26 million in 2012 from the Chesapeake Energy Corporation. This money was used to fund its campaign titled "Beyond Coal," advocating the use of natural gas in the utility industry. The irony lies in the fact that this funding period coincided with the time when residents of New York and Pennsylvania were protesting against the same company to stop fracking in their states (Davis, 2012b).

There also exist other factors like the fiscal health of a state, educational attainment of its population, and entrepreneurial leadership (Davis, 2012a) that exert considerable influence on states' decisions to allow fracking and regulate. Examination of each is beyond the scope of this chapter. Instead the discussion in this chapter has been limited to economic and political factors followed by state and local governments' responses to fracking. To answer the question posed at the beginning of the study (what are the existing state regulations on fracking?), attention has been drawn to the state of Texas' regulatory policy tools and its regulatory mechanism.

Perspectives on Fracking

In the United States, only two states have passed a complete ban on fracking, while some have proposed a ban where a state decision is still pending, and the rest have adopted a favorable stance toward fracking. In trying to understand states' stances on fracking, economy and politics cannot be ignored as they continue to play an important role in states' decision whether to allow fracking or not.

An Economic Perspective

In most states, the decisions on energy exploration and development are based on their economic prospects, and there exist the tendencies to overweigh the economic factors in the decision-making process. Often it is the lure of revenue and creation of jobs that plays a major role in states' decisions to continue fracking. This is even true at the local government level where additional revenue from fracking helps the jurisdiction to decide whether or not to impose a fracking ban (Cheren, 2014). In Texas, the non-profit Texas Oil and Natural Gas Association has claimed that in 2014 the oil and gas industry contributed $15 billion in royalties and tax revenue and created jobs for 2.4 million people in the state in various localities from oil- and gas-related activities (North Texans for Natural

Gas, 2015). The industry also provides $4 billion per year for education that goes to the schools and public universities in the state.

The lucrative scenario coupled with the fear of economic losses play an important role in states' decisions to allow or ban fracking. With the majority of the 33 states deciding to permit this type of unconventional drilling, their decisions seem to be based more on intuitive judgments rather than on rational choice. Under such circumstances, the tenets of Prospect Theory (Kahneman & Tversky, 1979) can partly help us to understand why states' decision makers tend to be risk averse toward economic losses and more risk acceptant toward environmental pollution from fracking. Applying the principles of this theory, it can be assumed that in the process of decision making, states do tend to calculate their losses and gains from a reference point of status quo followed by assessment of respective outcomes. When the prospect of likely losses from a decline in revenue, employment, individuals' monetary savings, and a rise in gasoline prices loom large, then such losses tend to get overweighed in the decision-making process. On the other hand, when the prospects of likely gains in environmental quality and improvements in individuals' quality of life by strictly regulating fracking do not seem significant in the absence of a detailed cost-benefit analysis, they tend to be isolated and undervalued in the decision-making process.

Additionally, states' unwillingness to regulate fracking can be partly attributed to the predominance of the market in a free enterprise economy that sometimes acts as a prison. Contrary to expectations, all government and business interactions in a market are not always in search of a common good (Peretz, 1986). A government operating within the confines of a market system can adopt a lackadaisical attitude toward regulations when dictated by the market. With businesses displaying a dislike for restrictive commands and the government fearing the loss of lucrative businesses as a result of restrictive regulations, the government is less likely to impose such restrictions on business activities. Further, the prospects of job and revenue growths are easily employed to distract elected politicians in energy producing states from the environmental ills of fracking. Needless to say, millions of dollars are spent by the oil and gas industry in lobbying the state government every year. In 2015 and 2016, the energy industry spent $130 million and $132 million approximately in lobbying the U.S. government at various levels. Among the big funders, Exxon Mobil ranked number one followed by Koch Industries and Royal Dutch Shell (Opensecrets, 2015). With lobbyists reminding the government of the industry's contributions to the economy and hiding relevant information from less informed government officials (Lindblom, 1982), it becomes difficult for the government to make changes and introduce new regulations.

In states with abundant mineral resources like oil and natural gas, it is the prospect of revenue from their exploration and production along with other economic benefits that take precedence over environmental protection and public health concerns of individuals living close to exploration and production sites. In these states, community members have not hesitated to express their disappointment and concerns over existing state regulatory policies, which tend to be biased toward the oil and gas industry. Evidence of such a bias can be observed in the passage of House Bill (HB) 40 in Texas. This bill was passed in 2015 and quickly became a state law in response to a threat posed by the city of Denton's declaration of a ban on fracking in late 2014. A few weeks after the Texas bill was passed, the state of Oklahoma followed suit. In 2015, Oklahoma repealed a statute that previously authorized local bans on drilling by a municipality as well. A similar stance has also been taken by the state of Louisiana (Ritchie, 2016). The passage of bills that preempt the power of local governments is becoming increasingly common among states irrespective of their identities as Republican or Democrat majority states. For example, in the Democrat majority states of New Mexico and Colorado, the state governments have also declared the local governments' restrictive ordinances or bans on fracking to be illegal because of contradictions with state law (Frosch, 2016; Ritchie, 2014).

A Political Perspective

From a political perspective, a common perception is that Democrat leaders are more likely to be pro-environmental than their Republican counterparts. Though an elected Democrat leader is expected to take a tougher stance on environmental pollution, that has not always been the case with fracking. In the states of Colorado and California, where both the governors are Democrats, the governors refused to pass a statewide ban or support local bans on fracking, despite strong public support for it. Instead, they opted for enforcement of new regulations to control fracking rather than totally dismissing this lucrative drilling activity from the state. Regardless of the political affiliation of states, the economic factors of job creation, revenue from oil and gas drilling, assurances from the oil and gas industry and trade associations that the technology is safe with little or no environmental consequences, and energy security continue to play an important role in decision making. Since these are intricately linked with a state's economy, they convey the message that any likely ban on fracking would spell an economic disaster, a risk which few politicians would be willing to take if they plan to seek reelection. For any decision to restrict or ban fracking is expected to have a repercussion on the political stand-ings of elected officials.

Another important factor that influences states' decisions on fracking is the prevailing political culture. For instance, the states of Vermont and New York, where fracking is banned, have a moralistic culture in Elazar's (1994) scheme of state political culture identity. In this culture, the public and politicians regard politics as the means to deliver public good. A government is considered as "good" when governance is done by those people who possess the virtues of honesty, selflessness, and commitment to public welfare. In this political culture, the ban on fracking is akin to the delivery of a public good by a "good government." Here, elected officials share the same concerns and passion as the public in protection of the air, water, and other valuable resources of the states.

The existing differences in political culture among states partly help to explain why some of them continue to permit fracking despite public outcry and complaints over the deterioration in environmental quality and other externalities arising from the process. In these states, biases toward job growth and revenue often result in overlooking individuals' demands for more environmental protection from fracking. For example, the ban on fracking in Denton that led the state of Texas to preempt the authority of local governments to pass such a ban on fracking in the state, simply suggests the presence of an authoritarian political culture here. Elazar has identified Texas' political culture as a mix of individualistic and traditionalist cultures. In an individualist culture, the emphasis is on governmental initiatives to promote private businesses and widen their access to the market with minimal community intervention. In a traditionalist culture, the emphasis is on the building of a hierarchical society with people higher up the hierarchy exerting a dominant control in decision making while those not involved in governance are discouraged from having a say in governmental matters. The combination of such cultural characteristics helps us to understand why decision making in Texas and other southern states is dominated by the power and social elites there. The political culture favors the election of conservative Republican governors in these states who continue to support the oil and gas industry and are reluctant to introduce any new legislation (Davis, 2012a) that might harm the energy industry, while the local residents and environmental groups have a limited say in the states' regulatory decision-making process.

Another variation in culture exists in the states of Colorado and California. Here, like in other states, the preponderance of economic concerns over environmental ones has made it possible to frack coal bed methane rocks and shale reserves. The existing political culture partly helps to understand why these states chose to adopt moderate regulations rather than a statewide ban on fracking. Perhaps it is the combination of moralist and individualistic values in the state culture that has made it possible not only for a political constituency with environmental concerns to thrive but also elect Democrat governors who are sensitive to environmental issues (Davis, 2012a). In the robust regulatory climate of both these states, there is more than one agency to oversee the implementation of regulations, and legislations have been introduced to soften the blow of fracking on those individuals living close to the drilling sites.

State and Local Governments' Responses to Fracking

The fracking of shale rocks initially seemed to be a boon for many states when people had little or no idea about its negative externalities. Over the years, its proliferation into communities has made people realize some of its drawbacks, and this has brought about opposition to fracking from community members. To seek adequate protection from fracking, community members have organized protests to put pressure on the state and local governments for changes in existing regulations. With federal regulations on fracking being weak and states assuming the bulk of responsibility to regulate, the latter's responses have varied. The states' responses are partly a reflection of political culture, the importance they attach to oil and gas production in the economy, and their commitment to environmental protection. On the other hand, the local governments in response to individuals' concerns have adopted a stance on fracking that does not always mirror that of the home state.

State Governments and Fracking

The state of Vermont was the first in the nation to issue a ban on fracking. In May 2012, Vermont passed an outright ban on fracking. The ban was more symbolic in nature as the state had little or no oil or natural gas production. The state's ban was more of a wakeup call to people in other energy rich states to focus on the environmental impacts of fracking through aggressive drilling tactics (Gram, 2012). In response to such a call for environmental protection, the state of New York issued a statewide ban on fracking, three years after the Vermont ban. Unlike Vermont, the state of New York has significant reserves of natural gas. The state's Democrat governor, Andrew Cuomo, formalized the ban on fracking in June 2015. Prior to the passage of the statewide ban, there was an intensive investigation into the environmental effects of fracking that lasted for seven years. During the inquiry period, the state's Department of Environmental Conservation, while conducting the investigation, received 260,000 public comments in its environmental impact study. The majority of individuals who submitted comments favored either severe restrictions or a complete ban on fracking. As a result, the announcement of the statewide ban on fracking came as no surprise. People living in communities close to fracking sites rejoiced at the news, which drew the ire of the oil and gas industry. The industry claimed the state's decision to ban fracking was political in nature rather than being based on scientific facts; therefore, some landowners opposed this ban fearing the loss of royalty payments and revenue from lease (Esch, 2015).

The state of California, noted for its stringent environmental regulations, refused to pass a ban on fracking. Its Democrat governor, Jerry Brown, declined to sign a bill to stop fracking in the state despite strong support from the public and environmental groups. After reviewing the number of jobs fracking has helped to create in the low-income areas of Central Valley where most of the fracking takes place, the governor decided to regulate the activity rather than ban it (Richardson, 2015). Meanwhile in Florida, a bipartisan group of lawyers lent support to a proposed ban on fracking in the state after a bill in the state to regulate fracking that would nullify the local bans on fracking in the state was turned down. The proposed ban is expected to safeguard the drinking water supply of communities, protect the environment, save tourism in the state, and promote sustainable development. The oil and gas industry has opposed the proposed ban and claimed that no such action is necessary since the technology has already proven to be environmentally safe and delivers many economic benefits. If passed, the ban would pose a threat to energy security (Calhoun, 2017). Similar sentiments on fracking have been expressed in the state of Maryland. In Maryland, a bill to ban fracking was introduced in the state senate in February 2017, before the expiration of the moratorium on fracking in October 2017. Opponents of fracking have criticized fracking because of the health risks it poses to people and hazards from earthquakes (Dresser, 2017).

Local Governments and Fracking

Local governments usually tend to support fracking because of state politics and federal government's emphasis on energy independence. Leaving aside the economic benefits from fracking, some community residents have developed a dislike of this type of drilling procedure with the disruption to their quality of life. The has prompted the local governments to make attempts to impose a ban on fracking which have produced mixed outcomes. A few communities in various parts of the nation have been successful in keeping fracking at bay through the passage of bans and moratoriums, while others have failed in such endeavors.

In California, the state's failure to impose a statewide ban on fracking did not deter several cities from passing their own local bans, some of which have already been overturned. A case in point is that of San Diego. In 2016, a temporary moratorium on fracking off the coast of San Diego had to be withdrawn when an environmental assessment conducted by two federal agencies found no significant impacts. Such findings prompted the Center for Biological Diversity that filed the initial lawsuit resulting in a moratorium to continue their opposition by filing an additional lawsuit to put a complete end to fracking off the coast of California (Nikolewski, 2016). Other states that passed similar bans or imposed restrictions on local governments' abilities to pass ordinances to stop oil and gas exploration and development include Pennsylvania, North Carolina, New Mexico, Ohio, and others.

In Colorado, the City Council of Longmont passed a voter-approved fracking ban in 2011. The ban prompted the Colorado Oil and Gas

Conservation Commission (COGCC) to sue this home ruled municipality in 2012, and the Colorado Oil and Gas Association supported COGCC in their efforts to reverse this ban. At the time of the passage of the Longmont ban, the state had over 50,000 oil and gas wells and over 45,000 inactive wells. The high level of investment already made in oil and gas development and production prompted the state to adopt a favorable stance toward the oil and gas industry. The state's Supreme Court ruled that the Longmont's fracking ban conflicted with the state laws and would increase the overall cost of oil and gas production, reduce royalties, and cause a ripple effect within the state economy, amounting to a statewide ban on fracking. Consequently, the court declared the ban to be invalid and unenforceable. In 2016, the court preempted Longmont of its authority to pass a ban on fracking in the future. Such a reversal only spelled victory for the oil and gas industry (Supreme Court of Colorado, 2016).

A similar case can be observed in the city of Denton, Texas. The negative externalities from fracking prompted the community members to seek changes in the local ordinances to restrict fracking. Unsatisfied with the outcomes, the city residents proposed a ban on fracking. The proposed ban was adopted as a ballot measure by the city and voters approved the ban on fracking in 2014. Threatened by the citywide ban on fracking in Denton and the adoption of a similar stance by other local governments in the state, the Texas Oil and Gas Association and the Texas Land Office immediately filed lawsuits against the city of Denton on grounds of the unconstitutionality of this ban. The state intervened and reversed the ban with the passage of House Bill 40 in 2015. This bill preempted all local governments' authority to pass a ban on fracking in the state and required any local regulations on oil and gas operations to be commercially reasonable, that is, providing opportunities to operators to pursue their drilling and related energy development activities in a cost-efficient manner (Heinkel-Wolfe, 2015). The oil and gas industry lauded the state in lifting the ban on fracking. On the other hand, individuals negatively impacted by fracking realized that state intervention made things worse for them.

Regulatory Tools and Regulatory Mechanism of Texas

Laws passed by the state legislatures and regulations enforced by state administrative agencies play an important role in protecting both public and private interests in a state. In Texas, the exploration and production of oil and gas has been facilitated by various laws of the state. Some of these laws date back to the early days of oil exploration and development in the state (Ghoshray, 2012). Others were developed and defined much later due to the fugacious nature of fracking products. Since both laws and regulations determine the scope of fracking in the state, attention has been drawn to Texas' laws on land ownership and management of water resources and its regulatory mechanism and regulations, discussed in the following sections.

State Laws Influencing Fracking

Land Ownership

When it comes to land ownership, the purpose of land ownership rights is to promote allocative efficiency, bestow legality in ownership, enable the owner to reap many benefits associated with ownership, and allow owners to enter into transactions for leasing or selling land at free will. In Texas, land ownership or property rights are subject to rules of the split estate system, meaning the property rights here have been severed into two parts – subsurface (mineral) and surface rights. The subsurface ownership or mineral right is a dominant right in the state. It can either belong to a single owner who owns both the surface and subsurface rights of the land or to the mineral owner of the land, who had retained the subsurface ownership rights even after the selling off of the surface rights. In any land transaction, the sale deed usually contains details on the type of rights that are being transferred in the sale transaction and helps to identify who is the actual owner of mineral wealth. The subsurface mineral rights owner decides whether or not to develop the mineral wealth lying under the ground. If a single owner possesses both the surface and subsurface land ownership rights, then the single owner has to explicitly deal with the gas company in the signing of a lease to develop the mineral wealth. In those cases of split property ownership, where surface and subsurface property rights are shared by two distinct owners, permission must be obtained from both owners before any drilling can begin.

In the event of unanticipated surface damages that are not covered in the contractual agreement of the owner with both mineral and surface rights, a conflict can possibly arise between the lessor and the lessee. Such cases are usually settled either through a mutual settlement or in court. An example of such type of a court case is Primrose Operating Company v. Senn in 2005 where the surface damages caused by oil and salt water spills on the property were subject to adjudication. The lessee refused to compensate for the damages in the absence of contractual agreement. After reviewing the case, the court's jury ordered the company to pay the owner $2,110,000 for actual and punitive damages but the Court of Appeals reversed the jury's verdict and sided with the operator. It ruled that the remediation of damages was not economically feasible and even though the spills led to a loss in property value, the overall increase in the market value of the land since the time of spills required no compensation for damages to the property owner (McFarland, 2009).

In addition to the signing of a lease between the mineral owner and the oil and gas company, permission also needs to be sought from the surface owner for access to the surface of the land. Part of the surface property is used in placement and storage of machineries that will be utilized in extraction of oil and gas. As per the state law, a surface right owner who does not support fracking cannot stop a mineral rights owner from having access to the surface property. The surface property owner has to allow the mineral owner to use as much land as possible for drilling purposes (Rahm, 2011). Furthermore, the mineral owner or the oil and gas company are not required by state law to compensate the surface owner for the loss of use of surface land, the clearing of trees, or for the damage of fences or hay pastures as long as these damages are within the limits of expectations and not due to negligence by the oil and gas producers. However, to avoid tension with the surface land owners, the oil and gas companies often do offer a lump sum compensation to offset the losses from use of their land (Woods, 2010) and for the release from liability for damages. As per the surface damage statute, the oil and gas company has to provide a five-day notice to the surface owner prior to staking, surveying, and other land evaluations, as well as a 30 days' notice before starting drilling operations. Also, the company should negotiate a good faith agreement with the surface owner as failure to do so would require either a waiver from the property owner or the posting of a $2,000 bond. If land owners consider the bond amount insufficient to cover for damages, they can challenge the company in state district courts (Earthworks, 2016).

Once drilling operations begin, the process of fracking, which calls for both vertical and horizontal borings into the layers of sedimentary rocks, promotes the migration of oil and gas trapped in the pores of sedimentary rocks to the drilling site from the adjacent property. The fugitive oil and gas can be captured by the producer without any liability to adjacent property owners. This results not only in a loss of mineral wealth to the adjacent and unleased mineral owners but also amounts to taking with no compensation under the aegis of the Rule of Capture. In 2008, the case of Coastal v. Garza in Texas helped to establish that any drainage from a neighbor's property is not considered a trespass. Therefore, it requires no consent from the adjacent property owner. In other words, under the Rule of Capture, drainage is permissible in the state and there is no liability involved in the capture process. It is interesting to note that this Rule of Capture is the same one that was used in the early days of the wild west to capture game animals. Originally derived from the common law of England, the old Rule of Capture is still in use to capture underground resources like groundwater, oil, and gas that may have strayed from adjacent properties (Kurth, Mazzone, & Mendoza, 2012). The state defends this rule in oil and gas exploration and development on the grounds of public interest. Only in those cases of proven damage to surface property as a result of drainage can an injunction be sought by adjacent property owners for violation of ownership rights (Gradijan, 2012; Rule, 2012).

In addition to land ownership rights, correlative rights also play an important role in establishing the rights of owners of their share of mineral wealth. Correlative rights refer to each mineral or property owner's right to a fair share of oil and gas when their parcel(s) of land lies above a common pool of mineral wealth. To protect the mineral ownership rights of unleased property owners, the Texas Rail Road Commission (RRC) requires the application of a permit for each well before drilling operations can begin. The purpose of the permit system is to ensure that both vertical and horizontal wells are spaced properly and stay off the limits of adjacent properties to protect the correlative rights of mineral or property owners.

In the large swaths of sedimentary formations, where fracking takes place only in isolated spots because few landowners have signed a lease with producers while others either did not consent to fracking or are preserving their land for future use, oil and gas production is not cost efficient. To remedy this problem, many states require compulsory pooling. Under the pooling option, regulatory agencies have the authority to compel adjacent property owners to sign a lease with oil and gas companies for the efficient production of oil and gas and not only receive compensation in the form of royalty payments but also contribute toward the cost of the production process based on land proportion. Unlike the neighboring state of Oklahoma, compulsory pooling is not well developed, its application is limited and not used that much (Warren, 2014). The limitations in the compulsory pooling statute along with its narrow interpretation by the state do not protect the correlative rights of owners, only offering flexibility to producers in oil and gas development. In fact, the state's Rule of Capture along with Rule 37 exceptions tend to conflict with the correlative rights of adjacent mineral owners.

After the oil or gas is extracted from beneath the land, there arises the need for its transportation. Usually, oil or gas is transported from the drilling site either through pipelines or by vehicles. In Texas, private pipeline companies have been granted the right to eminent domain under a state statute. This statute allows them to lay their pipelines wherever they deem suitable in development of a route for the transportation of oil and gas. Even in interstate transportation of oil or gas, the federal government provides the pipeline companies with the right to eminent domain under the Natural Gas Act of 1938. Pipelines can also be used to transport natural gas outside the country. For example, plans exist to transport natural gas from Eagle Ford Shale to Mexico to power the country's utility plants. The right to eminent domain not only enables a pipeline company to traverse through private properties but also allows them to seize land from private landowners through the process of condemnation. In such a process, the owners are required to be compensated for the fair market value of land as per the rules established by the State v. Carpenter case of 1936 (Brennan & Peacock, 2010). Changes occurred in 2004 and the Texas Supreme Court made amendments to its compensation scheme following the Hubenak v. San Jacinto Gas Transmission Company case. Through legal proceedings, the case managed to establish that there exists no need for compensation to the owner at the fair market value of the confiscated land. Also, in any litigation involving condemnation, if the landowner wins the case, then the attorney and appraiser's legal fees will not be compensated (Fambrough, 2009).

Water Laws

Another aspect of fracking that calls for attention is its water consumption. Fracking is a water intensive activity requiring thousands of cubic meters of water for the drilling of wells. In examination of water use for fracking there exists much ambiguity. In the state, water has been classified into three broad categories – natural surface water, diffused surface water, and percolating groundwater.

Natural surface water is owned by the state while the latter two categories are owned by the surface estate owner. If diffused surface water strays into natural watercourse, then it becomes subject to state law. When it comes to groundwater, the state's law is not based on the reasonable use doctrine, which is followed by many other states and calls for the withdrawal of water to maximize social benefits and minimize harm. Instead, the Texas Supreme Court has adopted the English common law rule of ownership. It confers to the surface estate owner the absolute right to capture and sell groundwater from underneath the property. Once again, under the state's Rule of Capture, a surface owner can pump as much water as possible from under the ground, even if it is detrimental to the neighboring owner. In other words, capture of water from the neighbor's underground water reserve is permissible by law without any liabilities to the adjacent owner as long as the capture is done from within the limits of one's property boundary. Such allowances by the state's groundwater law have helped it to earn a reputation as the "law of the biggest pump" (Eoh, 2014).

Under the state's common law, there exist certain restrictions on the surface land owner's right to withdraw groundwater for wasteful and malicious purposes, to prevent subsidence of land through careless drilling, and to drill a slant well that crosses the boundary lines of the surface property owner. When it comes to oil and gas drilling operations, once again the mineral owner's right in the use of groundwater for extraction of oil and gas supersedes that of the surface owner. If the mineral owner signs a lease with an oil and gas company, the latter does not need to seek the permission of the surface owner to drill a well for groundwater as long as the water use falls within the reasonable amount required for the exploration and development of oil or gas wells.

Regulations and Regulatory Agencies

In Texas, oil and gas explorations started at the beginning of the twentieth century. During that time period, in the absence of well-defined federal regulations and amidst expectations of a state's duty to regulate the environment, the state gradually developed its own regulatory framework over time. With the expansion of oil- and gas-related activities in the state, complexities in oil and gas exploration and development became evident. These have brought to focus a myriad of issues that need to be addressed in a fair and equitable manner.

Broadly speaking, fracking in Texas is subject to two types of regulations, general and specific. The general regulations refer to existing regulations on oil and gas development in the state which include fracking. These regulations are implemented by two state agencies, Texas RRC and the Texas Commission on Environmental Quality (TCEQ). The RRC serves as the primary regulator of oil and gas operations in the state while the TCEQ is mainly responsible for regulating air quality and the use of surface water in fracking. So far, there exists only one issue-specific regulation on fracking in the state with reference to the use of chemicals in fracking, which comes under the oversight of the RRC.

Texas Rail Road Commission

The Texas RRC was established through legislation in 1891 as the state's oldest regulatory agency. Initially, it had the authority to regulate rates and operations of railroads, wharves, and terminals in the state. Later with the expansion of oil and gas exploration and drilling activities in the state, the RRC was given the additional responsibility of regulating the oil and gas industry in 1919. This led to the creation of the Oil and Gas division at the RRC. At the beginning, the regulatory authority of the RRC on the oil and gas industry was challenged by independent oil well operators when the agency tried to impose control on the production of oil. It took several years for the courts and the state's legislature to clarify the RRC's regulatory authority over the oil and gas industry and in matters of conservation of the state's natural resources, protection of correlative rights, and prevention of environmental pollution. It was not until 1930 that the RRC fully assumed its regulatory role over the oil and gas industry (Railroad Commission, 2016).

Since the 1930s, the RRC, as the primary regulator of oil and gas operators in the state, assumed regulatory responsibilities under both state and federal laws for the implementation of the Safe Drinking Water Act, RCRA, Clean Water Act, Pipeline Safety Act, and several others. The mission statements of the RRC are in alignment with its responsibilities, which are to serve as the steward of the state's natural resources and environment, ensure the safety of the community, and promote economic vitality in the state. The agency is administered by a chair and two commissioners, who are elected. The oil and gas companies are allowed to contribute to prospective commissioners' political campaigns. To better understand the functioning of the oil and gas industry, the RRC employs people with work experience in oil and gas companies. There seems to exist an open-door policy between the RRC and oil and gas companies. In the past, a former oil and gas executive and a coal executive have found jobs as executive director and general counsel at the RRC, while a former commissioner of the RRC has assumed the role of a lobbyist for the oil and gas industry. The agency defends its policy of hiring people from the oil and gas industry claiming that to regulate it needs people who understand the oil and gas industry (Price, 2015).

The regulatory jurisdiction of the oil and gas division of the RRC extends not only over the oil and gas wells in the state but also over people who are either engaged in drilling or own and operate oil and gas wells. The agency's oil and gas divisions have regulatory responsibilities over various activities of the oil and gas industry from the beginning to the end, ranging from the spacing and drilling of wells to their plugging when operation ceases.

In the spacing of wells, the Rule 37 exceptions of the RRC enable oil and gas producers to access the common pool of oil and gas in rural and urban areas by using their own criteria. Initially, Rule 37 exceptions applied only to vertical wells, but with the introduction of fracking which also involves horizontal drilling, a supplemental Rule 86 was developed to include horizontal wells. Under the exceptions, if producers can prove that wastage of resources can be prevented or confiscation of properties avoided, then they are granted permission to disregard the RRC's well spacing requirements. Instead, they can use their own calculations in the spacing of wells close to unleased adjacent properties as observed in Barnett Shale, Eagle Ford near the Fort Worth area in south central Texas, and Haynesville formations in the northeastern part of Texas. Here, oil and gas companies use their own well spacing requirements and horizontal drilling to capture the fugitive oil and gas that migrates from adjacent property. By state law, they have no liabilities to mineral right owners from whose property the products have escaped (Behrens, 2011).

The RRC issues permits to operators for drilling or deepening wells for fracking. In the process of drilling, to prevent the contamination of underground water, it regulates the casing, cementing, drilling, and completion requirements of wells by enforcing the state's administrative code (3.13). Through the enforcement of another administrative code (3.8), it also regulates the storage, transfer, and disposal of oil and gas wastes to prevent surface water pollution. In those areas where saline or brackish water is extracted from below the base line of usable water for use in fracking, the RRC also regulates the drilling of water wells through its permitting process. Further, the RRC is responsible for setting the requirements for the plugging of wells under Statewide Rule 14. The state legislature requires operators to post $2,500 bonds for the plugging of wells.

When the time is ripe to plug a dry or inactive well, an operator has to submit a notice to the RRC of its intention and also keep the surface owner informed. An approval from the RRC and adherence to its guidelines are essential in the plugging of any oil and gas wells in the state. Failure on the part of an operator to plug a well prompts the RRC to complete the process and forces the operator to compensate the cost incurred. In recent years, with the downward slide in oil prices, the number of abandoned wells in the state has drastically increased. In 2015 alone, 94 oil companies abandoned 1,584 wells in Texas. Since these unplugged wells' pipes and casings can deteriorate over time and pose a substantial hazard to drinking water, many landowners have expressed concerns. The RRC has stepped in and plugged 692 wells, expending $11.7 million from its own fund. In view of the current crisis, the Texas Sunset Commission, which oversees the functioning of the RRC, has recommended an increase in the posting of bond amounts to cover the cost of plugging, and the RRC chair has also requested financial help from the legislature. However, the oil and gas companies along with their lobbyists are reluctant to bring about a change in the status quo. They continue to believe that the current bond amount and the few pennies received from the sales tax imposed on every barrel of oil for the Oil and Gas Regulation and Cleanup Fund are adequate to cover the cost of plugging (Tomlinson, 2016).

Besides overseeing the drilling and plugging of oil and gas wells, the RRC is responsible for pipeline safety. Its pipeline division is responsible for overseeing the safety of the state's natural and liquefied gas pipelines and natural gas transmission lines. Since there exist no statutory or regulatory requirements for a permitting process, the agency has limited regulatory authority over pipelines in the state. An exception to this rule is when a pipeline carries hydrogen sulfide gas, noted for its toxicity at certain levels. To prevent any damage to pipelines and for public safety, the RRC has developed regulations restricting excavations around pipelines and tries to promote public safety and awareness through its outreach program.

In addition, the RRC regulates solid hazardous wastes (as per definition of the EPA) that are not exempt from federal regulations. These include waste arising from the drilling, operation, and plugging of wells, separation and treatment of produced fluids at operation sites or natural gas processing plants, pipeline transportation of oil and gas, and other mining related activities at oil and gas production sites. To regulate hazardous wastes, the RRC adopted the statewide rule 98 in 1995 that became effective the following year. The rule 98 was aimed at prevention of pollution of surface and subsurface waters and injury to life and property that might occur from the mishandling of hazardous wastes. This rule is as stringent as that of the federal government's RCRA. Also, to reduce the volume of different types of hazardous wastes generated at production facilities, the RRC offers a Waste Minimization Program. This program helps the oil and gas companies in many ways – it lets them save money from disposal fees by reducing waste volume, decreases compliance and potential liability concerns, enhances efficiency in operations, and participation in the program helps the oil and gas companies to improve public relations (Railroad Commission, 2016). Additionally, the state amended its recycling rules (statewide rule 8, chapter 4/subchapter B) in 2013. Recycling is not mandatory in the state, but it is encouraged in the oilfield to reduce water use. Under the amended rules, recycling permits have been streamlined and rules for authorized fluid recycling have been created (Tamest, 2016).

The RRC's fracking related waste management policy, despite its good intentions to protect the public from the hazards of the oil and gas industry, has not always received support from the public. For example, in the state's small town of Nordheim, located in the Eagle Ford Shale, the residents organized a protest in the state capital to halt the approval of a proposed 143-acre site to store waste that would include drilling bits, oil-based mud, fracking sand, and other oil wastes. Despite the protest, the RRC voted 3 to 0 to approve the site in 2016. In its approval process, the agency paid more attention to the evaluation report of the waste site's effect on groundwater and the San Antonio-based waste handling company's assurances that safeguards like the waste liners and the thick layer of clay soil at the selected site would prevent the contamination of the groundwater. Township residents' social concerns, such as the site's proximity to school, noxious odor, increase in truck traffic, and its impacts on local roads that were not evaluated, were overlooked in the agency's decision-making process (Malewitz, 2016).

There exists only one issue-specific regulation on fracking. It was introduced in 2011 with the passage of Texas House Bill 3328 and signed by then state governor Rick Perry. This law became effective the following year, and all wells that received a drilling permit after February 1, 2012 came under the jurisdiction of this law (Gradijan, 2012). It called for the disclosure of chemicals used in fracking by oil and gas companies but it also offered protection of trade secrets. Under the new state law (Statewide Rule 29), gas and oil well operators are required to report the amount of water used in fracking and the names of those non-trade secret chemicals used in fracturing fluid (mainly their trade name, maximum concentration, supplier, and purpose) to FracFocus, a voluntary reporting website that serves as a national chemical registry. Developed in 2011, FracFocus is managed by the Ground Water Protection Council and Interstate Oil and Gas Compact Commission, both committed to conservation and environmental protection. Initially, 37 companies reported to this website; five years later, 1,000 companies report to the website. Its success has prompted 23 states in the nation to require oil and gas companies to release their chemical data via FracFocus so that it can be easily accessed by the public.

However, there exist two important exceptions to the state's disclosure law. If an operator is not provided with information on chemicals used in the fracking fluid by the manufacturer, supplier, or the service company, then this operator is exempted from supplying chemical names to the

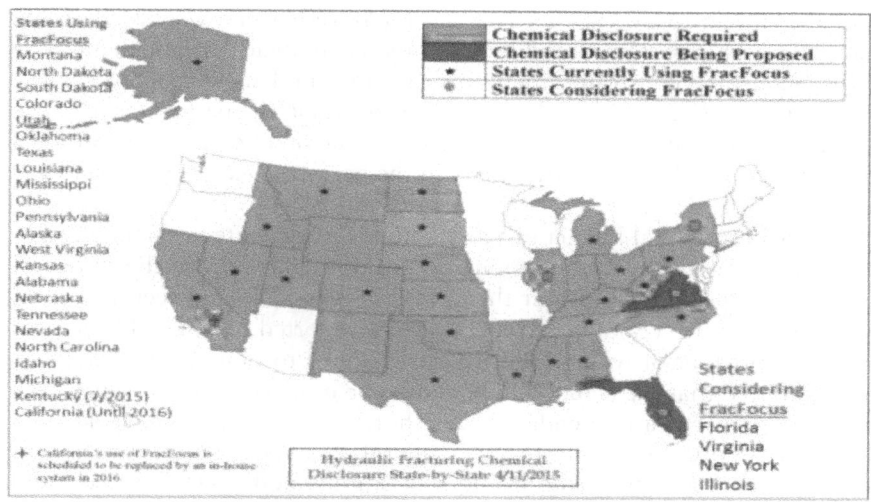

Figure 4.1 Types of Chemical Disclosure Requirements in the States of U.S.A.
Source: FracFocus, available at https://fracfocus.org/welcome.

national registry. In other circumstances, such as if chemicals that have been used in trace amounts, appear incidentally, or have not been intentionally added to the fracking fluid, then an operator on record is not required to reveal the names of those chemicals.

When it comes to protection of trade secrets, operators are required under the state's Government Code, Chapter 52, to file a claim in the Chemical Disclosure Registry form. Once the claim is processed and a protected status has been granted by the state government, operators are still required by law to provide the names of those trade secret chemicals to emergency responders and health professionals. The individuals receiving such information are required to treat the trade secret information as classified. Furthermore, a trade secret can be challenged in court by those who have been adversely affected by the chemicals. The landowner, adjacent property owner, the mineral owner, or a state agency can still challenge a trade secret within two years of the filing of a well completion report (Kulander, 2013).

Obstacles Encountered by RRC in Implementation of Regulations

Even with a well-developed scheme of regulations, the RRC has encountered some problems in implementation. It has an outdated computer system which makes it difficult to detect if wells and their owners have a past history of violations. The collected data is entered into a 40-year-old mainframe system, which limits the abilities of commissioners, employees, and the public to access the data if they have no knowledge regarding handling such a system. In addition to its technological woes, the agency faces budgetary constraints that obstruct the implementation of regulations. Since the agency's funding is dependent upon the performance of the oil and gas industry, that is, the fees paid by the oil and gas companies, its fiscal health is subject to fluctuations in the performance of the oil and gas industry in the state.

The RRC's annual budget of $87 million is inadequate in the implementation of regulations. With a population of 435,000 wells in the state and only 158 inspectors to inspect them, many wells have not been inspected for more than five years. In 2016, the agency reported issuing 1,400 notices of violations, out of which 700 involved failures to plug inactive wells. It also admitted that the actual number could be more since 65 percent of the wells have inspections pending. The shortfalls in funding have led the agency to petition the state for $45 million as a special appropriation to update its technology, make changes in storage and access of data, and hire more inspectors. But in the face of the enormity of the problem of backlogs, environmental group members have pointed out that even with more staff, it would still be difficult for the agency to clear the backlog, and 20 percent of the wells would still have a waiting time of more than five years (Handy, 2017).

Additionally, the RRC has been criticized for having a misleading name. Since the agency no longer deals with railroads, a name-change that is commensurate with its functions is necessary in the twenty-first century. The agency is aware of the dilemma but is yet to take any action.

Texas Commission on Environmental Quality

The TCEQ is another agency that has limited regulatory authority over the oil and gas industry in the state. Created in 1993, the TCEQ operates as a comprehensive Environmental Protection Agency. Since this agency resulted from the merger of the state's air, water, and waste regulatory programs, its mission is to ensure clean air and water in the state along with safe management of waste (TCEQ, 2016a). To control air pollution, the TCEQ requires the independent oil and gas operators to obtain permits from its office for authorizations on air emission from their facilities. Oil and gas production sites are also subject to Title V of the Federal Clean Act and the purchase of permits to operate. Additionally, air monitoring is conducted at various production sites in the Barnett and Eagle Ford Shale plays (TCEQ, 2016c). The data collected and the compliance history of facilities are saved and made available to the public as per state rule requirements from 2012 onwards. The Advanced Review of Compliance History, or ARCH, is a web-based program that reveals the names of violators and also gives the opportunity to operators to make necessary corrections prior to posting their annual compliance reports (Kulander, 2013), but it is available to registered users only.

Another function of the TCEQ is to regulate the use of surface water for fracking. At times of water shortages like droughts and emergency periods, the TCEQ director can suspend, modify, or interrupt the use of surface water. This suspension remains valid for at least 180 days and can become effective even without a prior notice. In addition to surface water, the TCEQ has regulatory control over reclaimed water. This type of water serves as an alternative source of water supply during periods of water shortages, but to use it in fracking, the operator has to secure a permit from the TCEQ. If the reclaimed water is from an industrial source, then no permit is required for fracking (TCEQ, 2016b).

Oil and gas activities produce much waste and its handling requires monitoring to prevent contamination of surface water and soil. The TCEQ is responsible for the management of waste generated at production sites, regulating over four types of oil and gas wastes. These include wastes from oil field service facilities like equipment and materials, waste generated during transfer of crude oil and natural gas by land and water modes of transportation, treated and processed wastes from oil and gas sites, and residential waste from the production site (TCEQ, 2016c).

Comparison of Texas' Regulations on Fracking with Other States

The regulatory framework of the state of Texas is more developed and detailed in comparison with some oil and gas producing states like North Dakota and Oklahoma. In those two states, oil and gas explorations and development are relatively recent activities and regulations are still in a nascent stage of development. Despite Texas' edge over other states in development of a more mature regulatory framework, there still exist deficiencies that need to be reviewed and addressed. A comparison chart (Figure 4.2) developed by the Public Citizen Texas, a non-partisan and public interest advocacy group, offers a glimpse into some of the strengths and deficiencies of the state's regulations by juxtaposing with those of other states in the nation, where fracking is also conducted.

The weaknesses in the state's regulatory mechanism and regulations can be traced to the lack of funding, transparency, and structural issues. Funding shortages have delayed the adoption of advanced computer technology and the hiring of inspectors responsible for enforcement of regulatory standards. The free information posted on agencies' websites is limited and sometimes lacks details. It is sometimes difficult to find up-to-date information and when found, the agency requires the payment of a fee for retrieval and release of the requested information. The Public Citizen Texas, an advocacy organization in the state, has pointed out that the RRC's website boasts statistics on the number of complaints on oil and gas companies received and resolved but there exist no other details. This

Public Citizen Texas

Oil and Gas Regulation/ Best Practices

			TEXAS	OKLAHOMA	COLORADO
Structure					
	Oil and Gas Regulatory Agency		Railroad Commission of Texas	Oklahoma Corporation Commission [1]	Colorado Department of Natural Resources: Oil and Gas Conservation Commission [2]
	Composition of Leadership		3 full-time elected commissioners: No bars to election based on conflicts of interest [5]	3 full-time elected commissioners: **must have *no direct or indirect interest* in any entity regulated by agency.** **Strict campaign limits, and *corporate contributions prohibited***, in addition, **contributions to *Commission candidates* limited to 120 days before primary and 120 days after general elections.** [6]	7 appointed part-time commissioners and 2 ex officio voting members, the Executive Directors of the Department of Natural Resources and the Department of Public Health and the Environment. **Specific requirements relating to education, experience, and location in state ensure *diversity and balance among appointed commissioners*.** [7]
	Jurisdiction/Functions of Agency		Primary jurisdiction over oil and gas industry, intrastate pipelines, gas services, including natural gas utilities and LP-gas industry, and surface mining operations. ***No jurisdiction over railroads.***	Regulates oil and gas activity, public utilities, and transportation, *including railroads*.	The DNR oversees forestry, mining, oil & gas, parks & wildlife, state lands, water conservation, and water resources
	Mission Statement		To serve Texas by our stewardship of natural resources and the environment, our ***concern*** for personal and community safety, and our ***support*** of enhanced development and economic vitality for the benefit of Texans.	In the interests of the public, the Commission will oversee the conservation of natural resources to avoid waste, abate pollution of the environment, and balance the rights and needs of the people with those of the regulated entities which provide essential and desirable services for the benefit of Oklahoma and its citizens.	To foster the responsible development of Colorado's oil and gas natural resources, which will result in the efficient exploration and production of oil and gas resources in a manner consistent with the protection of public health, safety and welfare; the prevention of waste; the protection of mineral owners' correlative rights; and the prevention and mitigation of adverse environmental impacts. **We are as committed to protecting public health and the environment as we are to fostering the responsible development of Colorado's oil and gas resources.**
Enforcement	Policy				Enforcement program is designed to deter violations and encourage compliance. 1. Notices of Alleged Violation [NOAVs] issued when Director has probable cause to believe violation has occurred. Less serious violations, i.e., violations that do not pose significant actual or threatened injury to public health or the environment, do not cause waste, do not damage correlative rights, *and* are not part of a pattern, may be handled with warning letter or corrective action required inspection report. 2. Most operators have opportunity to engage in settlement negotiations, resulting in consent orders. 3. If no settlement, matter proceeds to an Order Finding Violation [OFV] hearing, which is a full evidentiary hearing before the Commission. ****Violations that demonstrate an operator's gross negligence or knowing and willful misconduct, or pattern of violations must go directly to OFV hearing.**
	Inspections	***Alaska** has the lowest ratio with about **370** wells per inspector.	158 inspectors 134,484 inspections conducted in FY2015 Inspector-to-well ratio: **2,340+ active wells per inspector** **No inspection schedule for non-commercial disposal or injection wells**	The Field Operations Department has 50 field inspectors. The department is responsible for pollution prevention, insuring proper plugging of wells, witnessing mechanical integrity tests, running initial production tests on newly drilled wells, conducting UIC inspections and ensuring that good housekeeping practices are followed. The department is also responsible for identifying and prioritizing wells that need to be plugged using state funds. [19]	The Field Inspection Unit inspects oil and gas wells and related facilities. The Unit has 28 full-time employees Inspector-to-well ratio: 2,000 Wells with reported problems are prioritized for re-inspection, which may result in multiple inspections of the same well in a short period of time [20]
	Complainant Involvement		No role for landowners/complainants [22]		Complainants can track their complaints online, and can object to decisions of no violations and terms of proposed settlements. [23]

		WYOMING	NORTH DAKOTA	OHIO
Structure				
	Oil and Gas Regulatory Agency	Wyoming Oil and Gas Conservation Commission [3]	North Dakota Industrial Commission [4]	Ohio Department of Natural Resources
	Composition of Leadership	5 part-time commissioners: Governor, State Geologist, Director of Ofc of State Lands & Investments, and two appointed public members qualified to serve the oil and gas industry [8]	3 part-time commissioners: Governor, Attorney General, and Ag Commissioner [9]	Director is appointed by governor, cabinet-level executive position; no specific qualifications. The director designates the chief of the Oil & Gas Resources Division. **The chief shall not hold any other public office, nor be engaged in any occupation or business that might interfere with or be inconsistent with the duties as chief. [10]**
	Jurisdiction/Functions of Agency	Oil and gas activity [14]	Commission regulates certain utilities, industries, enterprises and business projects established by state law	Regulates all natural resources, including forestry, surface mining, oil and gas activity, parks and recreation, wildlife, soil and water conservation, energy, hunting/fishing, et al [15]
	Mission Statement		To encourage and promote the development, production, and utilization of oil and gas in the state in such a manner as will prevent waste, maximize economic recovery, and fully protect the correlative rights of all owners to the end that the landowners, the royalty owners, the producers, and the general public realize the greatest possible good from these vital natural resources.	To ensure a ***balance*** between wise use and protection of our natural resources for the benefit of all.
Enforcement	Policy			The Division works to protect Ohio's oil and gas resources, the environment and the interests of citizens living near oil and gas wells. Inspectors investigate citizens' complaints, enforce and oversee well construction and waste disposal activities, and the plugging of wells and site restoration. When a well owner fails to meet requirements established by law, the Division of Oil and Gas Resources Management has a variety of enforcement options to gain compliance. The Division generally maintains a standard operating procedure of escalating enforcement measures from informal to formal, depending upon the nature of the violation. When informal measures are unsuccessful or a violation endangers public health and safety or the owner demonstrates flagrant disregard for the law, more formal enforcement measures are used. The Division chief has the authority to issue orders or file civil or criminal enforcement actions, if necessary, to correct a violation.
	Inspections	Inspector-to-well ratio: more than 2,900 active wells per inspector.	18 inspectors Inspector-to-well ratio: **500**	ODNR has over 50 field inspectors, including four inspectors dedicated to inspecting Class II injection wells. **Injection wells are inspected by the Division once every 12 weeks, at a minimum**, but are often inspected more frequently. In addition to the quarterly inspections, the Division witnesses 100% of critical phases of the injection well construction and operation. This includes surface casing cementing, tubing and packer placement, and mechanical integrity testing.
	Complainant Involvement			

		WEST VIRGINIA	PENNSYLVANIA	NEW MEXICO
Structure				
	Oil and Gas Regulatory Agency	West Virginia Department of Environmental Protection	Pennsylvania Department of Environmental Protection	Energy, Minerals & Natural Resources Department
	Composition of Leadership	Governor appoints Secretary/Director [11]	Secretary appointed by Governor [12]	Governor appoints Cabinet Secretary [13]
	Jurisdiction/Functions of Agency	Air quality, mining, oil and gas activity, water use and quality, wastem dam safety, forestry	Primary responsibility for protection of natural resources and environmental health and safety	Department has jurisdiction over energy conservation and management, state forestry, mining and minerals, oil conservation, and state parks.
	Mission Statement	To support a healthy environment. Legislative finding: Those functions of government which regulate the environment should be consolidated...to carry out the environmental functions of government in the most efficient and cost effective manner, to protect human health and safety and, to the greatest degree practicable, to prevent injury to plant, animal and aquatic life, improve and maintain the quality of life of our citizens, and promote economic development consistent with environmental goals and standards. [16]	To protect Pennsylvania's air, land and water from pollution and to provide for the health and safety of its citizens through a cleaner environment. DEP's Office of Oil and Gas Management is responsible for the statewide oil and gas conservation and environmental programs to facilitate the safe exploration, development, recovery of Pennsylvania's oil and gas reservoirs in a manner that will protect the commonwealth's natural resources and the environment.	
Enforcement	Policy	Upon discovery of violation, the inspector shall write the violation with compliance requirement and provide a maximum of 7-day period to abate the violation. If the violation is an imminent danger, the inspector shall issue an imminent danger violation requiring the operator to cease operations. After the 7-day period, if the violation has not been abated, the inspector shall issue a failure to abate notice requiring the operator to cease further operations. Upon the issuance of a failure to abate, the inspector shall notify the Office's permitting section to block the issuance of permits. [17]	The primary objective of the enforcement program is to attain and maintain a high degree of compliance with the laws governing oil and gas development. **Basic Principles of Enforcement** - An appropriate enforcement action should be taken for each identified violation. - The minimum action for any violation is a written notification in the form of a NOV or a copy of an inspection report that notifies the operator or other responsible person (operator) of the violation. If the violation is corrected before the end of an inspection, an NOV may not be necessary if the violation is noted on the inspection report. - In order to attain and maintain compliance, follow-up action should be based on progressive enforcement in accordance with enforcement priorities of the program, as appropriate under the circumstances. - Civil penalties may be assessed in various circumstances. These include a threat to public health, safety or the environment; repeat violations; failure to apply for and obtain permits or registrations; and failure to submit reports. The factors used to determine the amount of the penalty include deterrence, damage to public health, safety or the environment; willfulness; operational cost savings; and costs incurred by the Department during an investigation.	Upon discovery of violation, an immediate enforcement may be taken (spill, pipeline rupture or blowout). Otherwise, a directive is first issued by the division to violators, If no response is received, or if a response is received which is inadequate, a Notice of Violation ("NOV") should immediately be issued and further enforcement action considered. If the response is adequate, a corrective action plan may be agreed upon. If no response is received to the NOV or if an inadequate response is received, the District or Bureau may make additional follow-up efforts to obtain voluntary compliance. An order may be issued shutting-in production of a particular well, unit or project, temporarily canceling oil and gas transport authority, ordering temporary abandonment, ordering permanent abandonment or suspending action on pending applications. If above measures failed, a decision on further enforcement action will be made. [18]
	Inspections	21 inspectors	DEP's Office of Oil and Gas Management had 100 inspectors in 2014; 26,940 inspections were conducted. [21]	47,399 inspections conducted in FY2015
	Complainant Involvement	Complainants have a role in certain investigation and enforcement matters.		

Figure 4.2 Oil and Gas Regulation/Best Practices.

Oil and Gas Regulation/ Best Practices

	TEXAS	OKLAHOMA	COLORADO	WYOMING	NORTH DAKOTA	OHIO	WEST VIRGINIA	PENNSYLVANIA	NEW MEXICO
Complaint Statistics available online	Total number of oil and gas complaints received and resolved; no details provided [24]	Searchable databases for well information, case processing, etc., but not very user friendly	Detailed reports, updated monthly, of information posted online; type of complaint, receipt and contact, days to resolution [25]		Daily activity reports (permit, producing well completion), annual production report	Public database of effective corrections made for violations [26]	Public website allows for broad or narrow searches; can search by operator, county, inspector, and/or timeframe. [27]	Public databases for inspections, violations, enforcement history, etc., searchable by operator or region. Compliance reports also available.	Detailed information about each well's depth, last inspection date, pit use status, financial assurance, compliance, complaints, incidents and spills, production/injection volume, points of disposition etc. Update nightly [28]
Penalties	Administrative penalty: I. up to $10,000 per day for violations not related to pipeline safety II. up to $200,000/day for violations related to pipeline safety. Criminal Penalty: I. Imprisonment of not less than two years or more than five years or a fine of not more than $10,000 or both for false applications, reports, and documents and tampering with gauges (V.T.C.A. Natural Resources Code § 91.143) II. Class A misdemeanor for a violation of disposal pit rule 91.452 (prohibited use of a saltwater disposal pit for storage or evaporation of oil field brines) and rule 91.457 (removal of unauthorized pit) III. not more than $10,000 per violation per day for violation of Natural Resources Code § 91.101 (Rules and Orders made to prevent pollution) [29]	Flat rate penalties ranging between $1,000 and $5,000. Criminal penalty: up to $5,000 and/or imprisonment for a term not exceeding 30 days [30]	Civil penalty up to the maximum of $15,000 per violation per day based on degree of violation; possibility of $1,000 per day penalty if threatened or actual impact to public health, safety, welfare, the environment, or wildlife. [31]	1. a fine up to $5,000 per violation per day 2. civil penalty up to $10,000 per violation per day 3. criminal penalty subject to a fine of not more than $5,000.00 and/or imprisonment up to 6 months [32]	1. Civil penalty up to $12,500 per day for each offense unless the penalty for the violation is otherwise specifically provided for and made exclusive 2. Class C felony if willfully violate rules pertaining to the prevention or control of pollution or waste [33]	1. violation of drilling permit: civil penalty up to $10,000 per day 2. violation of underground injection permit: civil penalty from $5,000 to $20,000 per day 3. wastewater discharge violations: civil penalty from $2,500 to $10,000 per day [34]	Penalties up to $2,500 per day per violation. **Horizontal wells, Civil penalties up to $5,000 per day per violation.** Civil penalties up to $5,000 and/or imprisonment up to 12 months [35]	**Conventional wells, civil penalty up to $25,000 per day plus $1,000 per day for each day during which the violation continues.** **Unconventional wells: $25,000 plus $5,000 for each day the violation continues.** [36]	1. Civil Penalty: up to $1,000 per day per violation. 2.Criminal Penalty: up to $1,000 per day and violation report not exceeding $5,000per day per violation not exceeding 3 years [37]
Hearings	Railroad Commission	OCC Office of Administrative Proceedings (separate division within agency)	COGCC , Hearings Unit [38]	Wyoming Oil and Gas Conservation Commission [39]	Oil and Gas Division	ODNR Division of Oil and Gas Resources	Oil and Gas Conservation Commission [41]	Pennsylvania Environmental Hearing Board [42]	Oil Conservation Division [43]
Presiding Officer(s)	In-house hearing examiners, legal and technical; may be conducted by commissioners, a director, or employees designated as examiners [44]	Division ALJs [45]	In-house hearing officers [46]	In-house examiners appointed by the commission or the full Commission [47]	Assistant AG along with OGD's technical staff [48]	Chief of the Division. "Any person adversely affected by an order by the chief of the division of oil and gas resources management may appeal to the oil and gas commission for an order vacating or modifying the order. (R.C. § 1509.36). 1.5 Appointed Commissioners (appointed by Governor) 2.No more than 3 members may belong to the same political party. 3.One member shall be classed as a representative of the public. 4.One member shall be classed as a representative of independent petroleum operations. 5.One member shall be classed as a representative of major petroleum companies. 6.One member shall be learned and experienced in geology.	Five Commissioners, including the DEP director and the chief of the Office of Oil and gas. The remaining three commissioners are appointed by the governor: one of them must be an independent producer and at least one must be a public member not engaged in an activity under the jurisdiction of the public service commission or the federal energy regulatory commission. The third appointee shall possess a degree from an accredited college or university in petroleum engineering or geology and must be a registered professional engineer with particular knowledge and experience in the oil and gas industry and shall serve as commissioner and as chair of the commission. [49]	Five appointed judges Qualification: one operator to tenure: 1. partner in a private law firm; 2. Assistant Counsel in DEP 3. Deputy Chief Counsel in DEP 4. partner in a private law firm; 5. presented case before DEP, worked inside DEP [50]	Division Examiner conducts hearing and makes recommendations; Director issues decision, which may be appealed to the Commission. [51]
Public Access to Information	Despite funding to improve technology, there is still very limited information available on the Commission's website	Searchable databases for well information, case processing, etc., but not very user friendly	Easily searchable databases and wealth of information available online: inspection/incident inquiries; facility inquiries; spill data, updated monthly; spill analysis by year; water-well data, updated monthly; field inspection reports; quarterly and annual enforcement reports		Daily activity reports (permit, producing well completion), annual production report	Public database of effective corrections made for violations	Information regarding complaints, investigations, and enforcement matters readily accessible on website	The public can search for individual permits (authorizations), operators (clients), wells (facilities), inspections, violations, and Program (DOG, oil and gas production information, permits issued, drilling commence date (DPCD data), county data, operator specific data, as well as inspections, violations and enforcement actions.) [52]	Detailed information about each well's depth, last inspection date, pit use status, financial assurance, compliance, complaints, incidents and spills, production/injection volume, points of disposition etc. Update nightly
Permitting	$200-300 based on depth [53]	$175 [54]	$0 [55]	$50 [56]	100 the permit expires one year after issuance [57]	• Non-Urban Drilling Permit $500.00 • Urban Drilling Permit $500-1,000 depending on population [58]	Conventional well: $400 Horizontal well: $10,000 [59]	Conventional wells: $250-$1,950+ (Based on well bore length) Unconventional wells: $4,200-55,000 (fee based on length) [60]	Search deef and locate statute, regulations, or policy addressing this issue
Drilling Fees	$300 per well [61]	$0 [63]	$0 [63]	$0 [63]					
Disposal Well Fees	$1,000 for commercial disposal well application. $300 for non-commercial injection or disposal well application [62]			$75 annual fee for all new and old disposal wells [64]	100 the permit expires one year after issuance [65]	Brine Disposal Permit $1,000 [66]	UIC permit fees: $550 (includes fee for groundwater protection plan) [67]	The same as production wells, above [68]	
Requirements			In addition to a drilling permit [review of subsurface issues], operators must also obtain a location assessment permit, which sets: operator identity; location, planned disturbance sizes, planned location; facilities, land use, soil and vegetation, surface and groundwater, habitat, cultural concerns, public safety aspects, and the operator's best management practices for construction and operation. [69]						

Oil and Gas Regulation/ Best Practices

		TEXAS	OKLAHOMA	COLORADO	WYOMING	NORTH DAKOTA	OHIO	WEST VIRGINIA	PENNSYLVANIA	NEW MEXICO
Financial Assurance	Bond per well	$2/ft of actual well depth, despite Commission figures indicating the average well plugging cost in FY 2015 ranged from $5 to $17 per foot. [70]	Based on plugging cost but not to exceed $25,000. The average cost of plugging a well over the last five fiscal years was $17,566, during the first six months of FY 2016 the average cost to plug a well was $2,940. Operators must submit an Affidavit of Well Plugging Costs to OCC. The minimum acceptable amount for OCC is $2/ft on the total depth of well. [71]	Plugging bond: $50,000 for wells less than 3,000 feet in depth, $20,000 for wells equal to or greater than 3,000 feet in depth [74] Surface bond for wells and associated facilities for which the surface owner neither owns the minerals nor has a surface use agreement with the operator - provides monetary award to surface owner for unreasonable crop loss or damage that cannot be remediated, and - requires $2,000-$5,000 individual well bond, or a $25,000 state-wide blanket bond. Agency has power to award $5 above bond amount. Additional bonds required for seismic operations, waste management facilities, inactive wells, UIC wells, etc. All operators are required to maintain general liability insurance of $1,000,000 per occurrence to cover property damage and bodily injury to third parties. [72]	$10 per foot, adjusted every three years based on the number of wells that may be drilled or in a lesser amount if approved by the director [74]	$50,000 (Wells drilled to a total depth of less than two thousand feet may be bonded in a lesser amount if approved by the director) [74]	$15,000 [75]	$5,000/conventional well $50,000/horizontal well [76]	$2,500 plus surcharges added to drilling permits to help fund abandoned well plugging program [77]	$50000 plus $1 per foot in some counties or $10,000 plus $1 per foot in others [77]
	Blanket bond	$25,000 to $250,000, based on the number of wells secured a. 1 to 10 wells, $25,000 b. 11 to 99 wells, $50,000 c. 100 wells or more, $250,000 [78]	$25,000 to $100,000.00 iLor prior operator's total net worth net less than $50,000 located in the state [79]	Plugging bond: $60,000 for less than 100 wells, $100,000 for more than 100 wells [80]	$100,000 [81]	$100,000 [82]		$50,000/conventional wells $250,000/horizontal wells [83]	$25,000 [84]	$50,000
Flaring/ Air Quality	Regulatory Agency	Railroad Commission	Oklahoma Corporation Commission	Colorado Department of Public Health and Environment and Oil and Gas Conservation Commission	Wyoming Oil and Gas Conservation Commission [85]	North Dakota Department of Health & Industrial Commission [86]	ODNR Division of Oil and Gas Resource	Department of Environmental Protection, Division of Air Quality [86]	Department of Environmental Protection, Bureau of Air Quality [87]	Energy, Minerals and Natural Resources Department, Oil Conservation Division
	Flaring Rules	*In Alaska, gas is prohibited from being released, burned or escaped into an except for safety or testing operation. Flaring or venting time not to exceed one hour.	Operators can flare up to 50 mcf per day without permit [88]	First state to adopt statewide rules to control venting and leaks from natural gas operations. Operators must capture 95% of gas. Operators must maintain ... prohibited. Rule 912.a Prior written approval required for flaring other than during upset conditions, well maintenance, well stimulation flowback, purging operations, or a productivity test. Rule 912.b. [90]	Up to sixty (60) mcf of gas per well per day can be vented or flared [91]	Operators can flare gas for one year from the date of first production from the well free of taxes, royalties, and penalties, and extensions can be granted for economic infeasibility. [92]	In unleashed areas where flaring is expected, the permittee must notify the local emergency response officials that such may occur. 1. Owner must use every reasonable precaution to prevent waste. 2. Gas may not be vented to atmosphere, must be flared if there is no economic market. [93]	'Temporary' flaring allowed for 30-days per year [94]	[95]	Operator may flare or vent casinghead gas produced from a well up to 60 days following completion; possible to apply for exemptions. [96]
Water Use	Regulatory agency	Texas Commission on Environmental Quality (RRC has no authority to regulate withdrawal or use of water)	Oklahoma Water Resource Board [97]	The Division of Water Resources	State Engineer's Office	State Water Commission [98]	ODNR Division of Soil and Water Resources	West Virginia DEP Division of Water and Waste Management	Pennsylvania DEP; Delaware River Basin Commission(DRBC); Susquehanna River Basin Commission (SRBC) [99]	State Office of Engineering [100]
	Permit Requirements	Permit required for surface water use only, groundwater belongs to surface owner. [101]	A 90-day provisional temporary permit required. [102]	No prior permit required, but must use water permitted for industrial use	1. a permit is required for both groundwater and surface water use. 2.Groundwater applications for projects over 25 gallons per minute, within a groundwater control area, must be approved by the control area's advisory board. In these control areas, an application also must be posted in a local newspaper. [103]	Permit required, each permitted water user is allocated a specific volume of permitted water use on annual basis and must report actual usage to the state. [104]	1. Water withdrawal registration required if facility has the capacity to withdraw 100,000 gallons per day. 2. Permit required if facility uses more than 2 million gallons of water per day [104]	Water use greater than 300,000 gallons to hydroffract a well must be reported. Horizontal well permit applications must include a water management plan if more than 210,000 gallons of water/30-day period will be used in conjunction with drilling, fracturing, or stimulating a well; Department determines suitability for permit based on types of water sources and disposal methods proposed. [105]	Operation must obtain a permit from SOE. From the permit is issued, proving the intended beneficial use of the water. The state engineer then issues a license to appropriate water to the extent and under the conditions of the actual application. [107]	
	Reporting/Monitoring	Operators required to report volume of water used in drilling/completion		Operators must report volume of water or fracking fluid used [106]		The permittee is required to submit a notice of commencement and a notice of completion with the State Engineer's office.	1.on-site remote telemetry to collect real-time data of water usage 2.Civil penalty up to $25,000 per day and criminal penalty [109]			Water masters in the Water Resources Allocation Program inventory water resources and monitor water use. [110]
Water Quality	Legal recourse if oil & gas activities result in diminished or disrupted water quantity to water rights owner	No			Yes	Yes	Yes	Yes	Yes	Yes
	Baseline water testing required	No	Yes, for injection wells [water wells]	Yes [water wells]	Yes [water wells and springs]	No	Yes	Yes [water wells and springs]	No, but encouraged because may provide defense against presumption of liability.	No
	Presumption of liability for pollution found in nearby water source	No	No	No	No	No	No	Yes	Yes	No

Public Citizen Texas

Oil and Gas Regulation/ Best Practices

		Texas	Oklahoma	Colorado	Wyoming	North Dakota	Ohio	West Virginia	Pennsylvania	New Mexico
	Regulatory Agency	Railroad Commission	Oil and Gas Conservation Division	Colorado Oil and Gas Conservation Commission	Wyoming Oil and Gas Conservation Commission [111]	Industrial Commission and Department of Health	ODNR Division of Oil and Gas Resources	Office of Oil and Gas	Department of Environmental Protection	Oil Conservation Division
Waste	Fracking Fluid	1.recycling or reuse encouraged 2.advised underground injection [112]	Recycle or underground injection	1.Recycling or reuse 2.Underground injection 3.Disposal at a commercial solid waste disposal facility 4.Land treatment or land application at a centralized E&P (exploration and production) waste management facility [113]	1.underground disposal 2.reuse	All the waste materials of exploration and production must be disposed of in an authorized facility [114]	1.Disposal by injection at a permitted Class II well (88 percent) 2.Recycling of flowback is not typically done 3.Nearly two percent is spread for dust and ice control. 4.A permit is required [115]	1.Reuse 2.Underground injection 3.Transport to other state	1. treatment and discharge to surface water 2.underground injection 3. reuse 4. transportation to out-of-state facilities [116]	1. recycle, reuse 2. underground injection 3. discharge to surface water [117]
	Pit Use	For temporary storage. Permit is sometimes not required (for Admin. Code tit. 16, § 3.8 d. 4.)	For temporary storage	Temporary and permanent	Temporary and permanent; permit required for all	Yes, for evaporation or for further disposal arrangements [118]	Yes, for temporary storage	Yes, for temporary storage	New rules eliminate the use of waste storage pits for unconventional operation [119]	Temporary and permanent pits: no permit required for closed-loop systems. • "Permanent pit" means a pit used for drilling... [120]
	Liner requirement	All pits must be lined. The liner shall be designed, constructed, and installed to prevent any migration of materials from the pit into adjacent subsurface soils, groundwater, or surface water at any time during the life of the pit. [121]	Concrete, steel, or geomembrane liner requirement [122]	Liners generally required; earthen pits allowed for some uses.	Requirements vary depending on conditions at the site and the pit's use.	Sufficient impermeable [123]	1.Synthetic liner pursuant to the division's specifications 2.All pits used for the temporary storage of saltwater and oil field wastes shall be lined with liquid-tight [124]	An impermeable synthetic liner requirement except when the soil is deemed to prevent seepage, leakage, and overflows [125]	New Rule: The unconventional industry will be prohibited from utilizing pits to store drill cuttings and waste fluids. [126]	1. Geomembrane liners required for temporary pits 2. Each permanent pit must contain, at a minimum, a primary (upper) liner and a secondary (lower) liner with a leak detection system appropriate to the site's conditions; both must be geomembrane liners. [127]
	Closure time	All completion/workover pits used when completing or working over a well shall be dewatered within 30 days and backfilled and compacted within 120 days of well completion. All completion/workover pits used when working over a well shall be dewatered within 30 days and backfilled and compacted within 120 days of completion of workover operations. [128]	Pit shall be closed within six months or twelve months after drilling has ceased. Closure time is different for different areas. [129]	Unless an extension approved by Director, a pit can be used for no more than 3 years for drilling or reuse, treatment, or disposal of E&P waste or fresh water [130]	Site reclamation must be initiated within one year of permanent abandonment of a well or last use of a pit. [131]	The contents of the pit or receptacle must be removed within seventy-two hours after operations have ceased. Earthen pit should be reclaimed within 30 days after operations have ceased. An extension may be granted by Director up to one year. [132]	2 months after the date drilling is completed in non-urban areas and 14 days in urban areas the well owner/agent shall empty and fill in all pits [133]	Six months after the completion of the drilling process [134]		1. An operator shall close a permitted permanent pit within 60 days of cessation of operation of the pit. 2. A temporary pits must be closed within six months from the date that the operator releases the drilling or workover rig. [135]
Injection Wells/ Induced Seismicity	Permit requirements	A permit for a Class II disposal well may be modified, suspended, or terminated if injection is likely to be or determined to be contributing to seismic activity. * Completely discretionary and no guidelines. [136]	Each well requires notice and hearing before permit issued unless within an area permit that previously had public hearing. Traffic Light System for risk management: "yellow light" wells that meet "Red light" standard shall be shut down. In the event of increased earthquakes in area. For "yellow light" wells, the injection permit is temporary and wells must shut down every 60 days and bottom hole pressure readings taken [137]	Safeguards in permitting process to reduce likelihood of induced seismicity: limits on injection volume and rate, and requiring maximum allowable injection pressure to be set below the fracturing pressure for the set injection zone. Permit application review also includes evaluation of area for seismic activity [138]	The average and maximum disposal pressure. The estimated minimum and maximum amount of water to be injected daily [139]	When applying for a permit for class II UIC wells, operators need to provide the agency with information including "the estimated bottom hole fracture pressure of the top confining zone" and "average and maximum surface injection pressure." [140]	Operators must run a complete suite of geophysical logs on newly drilled Class II disposal wells. Installation of monitoring technologies required for new Class II application permits, including a continuous pressure monitoring system and automatic shutoff system. Also toughen permit requirements for drilling activities near faults and areas of seismic activity ... Additional permit requirements for Class II disposal wells on a web-by-well basis, including ... minimum geophysical logging suite, radioactive tracer or spinner survey [141]	1. Conduct a detailed geologic investigation of subsurface features in the vicinity of the injection well. The investigation must assess for the presence of subsurface faults, fractures or potential seismically active features. 2.Operator shall provide proposed operating data: a. Average daily rate or volume of fluid to be injected b. Maximum daily rate or volume of fluid to be injected c. Average injection pressure d. Maximum injection pressure [142]	Although the state has not had earthquakes connected to fracking or deep wastewater injection wells, in addition to a well permit from the DEP, injection wells require a UIC permit. Injection wells ... since 2011 also have operating conditions requiring continuous monitoring for injection pressure. [143]	
	Monitoring/ Reporting	Monthly average injection rates, total monthly volumes, and maximum wellhead injection pressures for wells [143]	Proposed rules would require all disposal wells within an area of interest to record volumes and pressures daily, and report, at a minimum, weekly or as otherwise directed by the division. [144]	Injection volumes must be reported monthly [145]	Monthly reporting required of the type and source of the injected substances, the total amount injected, and the injected pressures and casing-tubing annulus pressure during injection. [146]	The average injection pressure must be reported monthly [147]	The owner/operator of a liquid injection or waste disposal well must monitor daily pressure, flow rate, and cumulative volume for produced fluid disposal operations. Recording of one observation of injection pressure, flow rate and cumulative volume at reasonable intervals not greater than 30 days. [148]		Observation weekly of injection pressure, flow rate, and cumulative volume for produced fluid operations: Recording of one observation of injection pressure, flow rate and cumulative volume at reasonable intervals not greater than 30 days. [149]	Monthly water disposal report [150]
	Pressure testing	Each disposal well, fluid injection well, and gas reservoir storage well must be pressure tested before injection operations begin and at least once every five years or more frequently if required by the Commission. MIT: some wells need annual testing [151]	The test must be witnessed by an authorized representative of the Conservation Division [152]	1.After operators obtain a disposal permit, a mandatory initial pressure test of casing tubing annulus must be passed before any injection. 2.For noncommercial disposal wells, subsequent mechanical integrity test (MIT) must be done once a year or once every 5 years based on injection volume. 3.for commercial disposal wells, subsequent MIT must be done once every 12 months. [153]	Maximum surface injection pressure is calculated based on a default fracture pressure gradient of 0.6 pounds per square foot ("psi") of depth. The injection pressure is uniquely defined for each well [153]	Operators must provide the opportunity to witness all integrity tests. The application for disposal of saltwater must provide the average and maximum disposal pressure [154]				

Public Citizen Texas

Oil and Gas Regulation/ Best Practices

	TEXAS	OKLAHOMA	COLORADO	WYOMING	SOUTH DAKOTA	OHIO	WEST VIRGINIA	PENNSYLVANIA	NEW MEXICO
Regulatory solutions	Essentially none. The website states that "[a]t this time, the Commission has not determined that regulatory activities are warranted..." They refuse to acknowledge the problem, and therefore, offer no meaningful solution. However the Legislature recently funded the TexNet Seismic Monitoring Program, through UT's Bureau of Economic Geology, to deploy additional sensors around the state, and clarify and determine whether increased seismic activity is linked to the underground disposal of oilfield waste.	OCC has developed an earthquake response plan including disposal wells shut-in, injection volume reduction, or injection depth reduction for Western and Central Oklahoma. Recent agreement with operators also gives researchers "with to study earthquake problem." Traffic Light System for risk management: wells that meet "Red light" standard shall be shut down in the area, for "yellow light" wells, the injection permit is temporary and wells must shut down every 60 days and bottom hole pressure readings taken [155]	Regulators ordered a temporary shutdown of an oil and gas injection well operation near Greeley in 2014 after seismologists detected two earthquakes in the area in less than a month. [156]		The Commission can modify or suspend the well permit at any time, in addition the Commission may order the operator to cease injection at a well should it become noncompliant. [157]	Prohibited all drilling into the Precambrian basement rock			The state Department of Conservation and Natural Resources and the Department of Environmental Protection said they will spend $551,000 on a network of seismic activity monitors at 30 stations across the state for three years. [158]
Policy	Homeowners can sue the oil and gas industry for injuries or property damage resulting from earthquakes. [159]		Pending legislation (HB 1310) would hold operators strictly liable for their conduct of oil and gas operations, including hydraulic fracturing, treatment or reinjection operation, cause an earthquake that damages property or injures an individual. [160]						
Tax Distribution/ Trust Fund		One percent of gross production (unconventional wells) returned to local government [161]	2. Approximately 49% of the rentals and royalties from federal lands are returned back to the state of origin	Local government property (ad valorem) tax/CNT production is assessed at 100 percent of the prior year's market value of production. Local tax levies are applied to the assessed value. Local property taxes on gross production tax are collected directly by local governments and distributed to counties, schools, cities, and special districts based on the location of production and local mill levies [162]	Local governments retain the first $5 million generated from the gross production tax, and 25 percent of additional revenue generated locally during each biennium [163]				Local Government Property (Ad Valorem) Tax. Local levies vary between 8 and 11 percent on taxable value [164]
Money distributed to local gov per unconventional well	$211,535 [165]		$1,121,586 [165]	$1,234,638	$338,853 [167]				$176,732 [168]
Disclosure of Chemicals		1. The supplier or service company shall provide to the operator of the well information concerning each chemical ingredient intentionally added to the hydraulic fracturing fluid not later than 15 days following the completion of hydraulic fracturing treatment(s) on a well. 2. The operators need to submit information to FracFocus on or before a well completion report is submitted to the Commission. 3. Trade secret exemption [169]	Within 60 days of completion, operators must submit information to FracFocus or the Commission: 1. Trade name, supplier, and general purpose of each additive. 2. Associated Chemical Abstract Service Numbers for each additive. 3. Total volume of "base fluid" used. [170]	Ingredients and concentrations must be posted on FracFocus website within 60 days of completion; trade secret protection but must include chemical family or other similar descriptor associated with claim. [171]	Chemical additives, compounds and concentrations or rates proposed to be mixed and injected must be reported prior to use. [172]	Must report all chemicals and the maximum concentration of each chemical to division within 60 days; trade secret exception. [173]	Must disclose chemicals/additives and total volume of fluids used to agency; trade secret exception.	Within 30 days after completion of the well, a completion report containing chemical additives must be reported to the Department. [175]	Very detailed description of the hydraulic fluid composition and concentration listing each ingredient, the maximum ingredient concentration in each additive, the maximum ingredient concentration in the hydraulic fracturing fluid must be disclosed within 45 days; trade secret exception. [176]
Miscellaneous	*Indicates additional, significant findings from other states.	*In Arkansas, flow meters, or other measuring devices approved by the Director, must be installed on all Class II Disposal and Class II Commercial Disposal Wells. Permit Holders must submit accurate injection volume and pressure information, on no less than a daily basis, on a form prescribed by the Director. A formula is used by the Director to determine the maximum permitted injected pressure.	The Oil and Gas Health Information and Response program was created to respond to public concerns about health related to oil and gas activities. Clearinghouse of Oil and Gas Health Information. COGC has been working to strengthen its oversight of oil and gas development in Colorado. Since 2011, together with government stakeholders, they have crafted rules to lengthen distances between wells and homes, neighborhoods, reduce the effects of light, noise and odors, protect groundwater, cut significantly reduce methane and other chemicals, increase spill reporting, toughened requirements for operating in floodplains. They have also significantly expanded oversight and increased resources of access and volume of data available to the public. COGC collaborated with local governments, sponsored independent studies to increase understanding of impacts to air and water and adopted several formal issues brought about by new technologies and increased energy development. In January 2016, Commission approved new rules that amplify the role of local governments in siting of certain...					Model procedure for meaningful public participation in permitting process. Under the guidance of the Environmental Justice Advisory Work Group, DEP also developed the EJ Enhanced Public Participation Policy. The policy was created to ensure that EJ communities have the opportunity to participate and be involved in a meaningful manner throughout the permitting process when companies propose permitted facilities in their neighborhood or when existing facilities expand their operations.	

[1] http://www.occeweb.com/

[2] http://cogcc.state.co.us/#/home

[3] http://wogcc.state.wy.us/

[4] https://www.dmr.nd.gov/oilgas/

[5] V.T.C.A., Natural Resources Code § 81.01013 Conflict of Interest. Only applies to employees

[6] Oklahoma Constitution, Article IX Section 15
http://www.blueprintsfordemocracy.org/model-state-contribution-limits-and-source-prohibi-tions/Oklahoma Ethics Law TITLE 17, Corporation Commission § 48.

[7] Colorado Oil And Gas Conservation CommissionTITLE 17, Corporation Commission § 48.

[7] Colorado Oil And Gas Conservation Commission 2014 Annual Report 6, available at https://www.colorado.gov/pacific/sites/default/files/SB181arCOGCC2014.pdf

[8] http://governor.wyo.gov/media/news-releases/2014-news-releases/wyomingoilandgascommissionnameswatsonpermanentsupervisor

[9] http://www.nd.gov/ndic/

[10] R.C. § 1501.05
R.C. § 1509.02

[11] http://www.dep.wv.gov/Documents/employeeaddress.pdf
page:76

[12] http://www.dep.wv.gov/oil-and-gas/Impoundments/Documents/Impoundment%20Refernces/WV%20Code%2035-04.pdf §35-4-2. 2.3
http://www.dgs.pa.gov/Documents/Vol%20121%20-%20Entire%20Manual.pdf
Page4-57

[13] http://law.justia.com/codes/new-mexico/2013/chapter-70/article-2/section-70-2-5

[14] W.S.1977 § 30-5-104

[15] Ohio R.C. § 1501.01, et.seq.

[16] W. Va. Code sec. 22-1-1(5)

[17] WEST VIRGINIA CODE §22-6-6.
http://www.dep.wv.gov/oil-and-gas/Resources/Policy/Documents/Enforcement%20Policy.pdf

[18] http://www.emnrd.state.nm.us/OCD/documents/Enforcement_guidelines.pdf

[19] http://www.occeweb.com/News/FY14%20ANNUAL.pdf
page 29

[20] http://cogcc.state.co.us/documents/about/TF_Summaries/GovTaskForceSummary_FieldInspectionUnit_Overview.pdf

[21] http://files.dep.state.pa.us/OilGas/BOGM/BOGMPortalFiles/Annual_Report/2014/2014_Annual_Report_for_web_July1.pdf

[22] http://www.rrc.state.tx.us/oil-gas/complaints/

[23] Rules 522.a.(4) and 503.b.(4;Rule 522.b.(4) and 503.b.(4)

[24] http://www.rrc.texas.gov/media/29356/fy-15-oil-and-gas-enforcement-data-revised-4th-qtr.pdf

[25] http://cogcc.state.co.us/documents/complaints/Complaint_Detailed_Report_2016_3_8_2016

[26] http://apps.ohiodnr.gov/oilgas/rbdmsreports/Reports_Corrections.aspx

[27] https://apps.dep.wv.gov/oog/svsearch_new.cfm?pageType=viol

[28] http://www.emnrd.state.nm.us/ocd/ocdpermitting/Data/WellDetails.aspx?api=30-025-41177

[29] 1. http://www.statutes.legis.state.tx.us/Docs/NR/htm/NR.81.htm Section 81.0531
2. V.T.C.A., Natural Resources Code § 85.381

[30] Criminal:
52 Okl.St.Ann. § 278
52 Okl.St.Ann. § 247

[31] 2 CCR 404-1:523 (c)

[32] W.S.1977§ 30-5-119

[33] NDCC 38-08-16

[34] http://codes.ohio.gov/orc/1509.33v1

[35] § 22-6-34. Offenses; penalties.
§ 22-6A-19. Offenses; civil and criminal penalties
http://www.legis.state.pa.us/cfdocs.legis/LI/consCheck.cfm?txtType=HTM&ttl=58

[36] 58 P.S.§ 3256. Civil Penalties
New Mexico Statute Chapter 70
70-2-31. Violations of the Oil and Gas Act; Penalties.

[37] https://cogcc.state.co.us/Hearings/HearingGuide.htm
http://wogcc.state.wy.us/

[38] NDCC 54-57-03

[39] http://www.legis.state.wv.us/wvcode/ChapterEntire.cfm?chap=22c&art=9

[40] §22C-9-4
§22C-9-10

[41] http://cfb.courtapps.com/content/2014AnnualReport.pdf
page 3

[42] http://www.emnrd.state.nm.us/OCD/hearings.html

[43] 16 TAC § 1.121

[44] https://www.occeweb.com/FY13%20Annual%20Report%20FOR%20PRINTING.pdf

[45]

page 13

[46] https://cogcc.state.co.us/Hearings/HearingGuide.htm
[47] http://soswy.state.wy.us/Rules/RULES/7930.pdf
[48] https://www.dmr.nd.gov/ndgs/documents/newsletter/2007Winter/OGhearing.pdf
[49] http://www.legis.state.wv.us/wvcode/ChapterEntire.cfm?chap=22c&art=9
§22C-9-4
Of the three members appointed by the governor, one shall be an independent producer and at least one shall be a public member not engaged in an activity under the jurisdiction of the public service commission or the federal energy regulatory commission. The third appointee shall possess a degree from an accredited college or university in petroleum engineering and geology and must be a registered professional engineer with particular knowledge and experience in the oil and gas industry and shall serve as commissioner and as chair of the commission.
[50] http://ehb.courtapps.com/content/Bios.php
http://www.legalspan.com/catalog2/faculty.asp?UserID=20040311247242281046%20%20%20%20%20%20%20&OwnerColor=%23003366&recID=20091026-150226-74531
http://www.legalspan.com/catalog2/faculty.asp?UserID=20040311247242775153%20%20%20%20%20%20%20&OwnerColor=%23003366&recID=20101025-150226-85254
http://www.legalspan.com/catalog2/faculty.asp?UserID=201301172291941226 59%20%20%20%20%20%20&OwnerColor=%23003366&recID=20130815-229194-145350
http://www.legalspan.com/catalog2/faculty.asp?UserID=20130117229194130524%20%20%20%20%20%20&OwnerColor=%23003366&recID=20121009-229194-104006
[51] http://www.emnrd.state.nm.us/OCD/hearings.html
[52] "http://www.depreportingservices.state.pa.us/ReportServer/Pages/ReportViewer.aspx?/Oil_Gas/OG_Compliance", " Department Data Information System.
[53] http://texreg.sos.state.tx.us/public/readtac$ext.
TacPage?sl=R&app=9&p_dir=&p_rloc=&p_tloc=&p_ploc=&pg=1&p_tac=&ti=16&pt=1&ch=3&rl=78
[54] http://www.occeweb.com/ad/FeeScheduleFY2013.pdf
[55] 2 CCR 404-1: Appendix III
[56] http://soswy.state.wy.us/Rules/RULES/7928.pdf
Section 8 (a)
43-02-03-16
[57] http://www.dmr.nd.gov/oilgas/rules/rulebook.pdf
[58] http://oilandgas.ohiodnr.gov/portals/oilgas/pdf/FEE-SCHEDULE.pdf
[59] W. Va. Code St. R. § 35-4-526; §35-8-5
[60] http://www.pacode.com/secure/data/025/chapter78/chap78toc.html
§ 78.19. Permit application fee schedule.
[61] http://texreg.sos.state.tx.us/public/readtac$ext.TacPage?sl=R&app=9&p_dir=&p_rloc=&p_tloc=&p_ploc=&pg=1&p_tac=&ti=16&pt=1&ch=3&rl=78 (b)(4)
[62] Okla. Admin. Code 165:5-3-1(b)(1)
[63] 2 CCR 404-1: Appendix III
[64] http://soswy.state.wy.us/Rules/RULES/7929.pdf
Section 5(a)
[65] http://logcc.ok.gov/Websites/logcc/images/2013_SOS/NorthDakota2012.pdf

page 5

[66] http://oilandgas.ohiodnr.gov/portals/oilgas/pdf/FEE-SCHEDULE.pdf
[67] http://www.dep.wv.gov/oil-and-gas/GI/Forms/Documents/UIC%20APPLICATION%20PACKAGE%2006-25-2014,pdf
[68] http://www.pacode.com/secure/data/025/chapter78/chap78toc.html
§ 78.11. Permit requirements
§ 78.15. Application requirements
§ 78.19. Permit application fee schedule.
[69] http://cogcc.state.co.us/documents/about/TF_Summaries/GovTaskForceSummary_Environmental_OGLA.pdf
[70] www.rrc.state.tx.us/oil-gas/compliance-enforcement/hb2259hb3134-inactive-well-requirements/cost-calculation
[71] http://www.occeweb.com/rules/CH10eff09-12-14searchable.pdf
165:10-1-12.(a)
http://okwnews.com/news/whatzup/whatzup-politics/111018-legislators-vote-to-extend-well-plugging-fund-5-more-years.html
ftp://occ.state.ok.us/OCCFILES/instruct.htm#1006DINST
Form 1006D
[72] https://www.sos.state.co.us/CCR/GenerateRulePdf.do?ruleVersionId=6438
rule 706
[73] WOGCC Rules
Chapter 3 Section 4 (b)
[74] https://www.dmr.nd.gov/oilgas/rules/rulebook.pdf
43-02-03-15. BOND AND TRANSFER OF WELLS.
[75] http://codes.ohio.gov/orc/1509.225v1
[76] W. Va. Code, § 22-6-26; §22-6A-15
[77] http://www.pacode.com/secure/data/025/chapter78/chap78toc.html
§ 78.303. Form, terms and conditions of the bond.
[78] http://texreg.sos.state.tx.us/public/readtac$ext.
TacPage?sl=T&app=9&p_dir=F&p_rloc=164960&p_tloc=14567&p_ploc=1&pg=2&p_tac=&ti=16&pt=1&ch=3&rl=78
[79] http://www.occeweb.com/rules/CH10eff09-12-14searchable.pdf
165:10-1-10. (a)
[80] https://www.sos.state.co.us/CCR/GenerateRulePdf.do?ruleVersionId=6438
rule 706
[81] WOGCC Rules
Chapter 3 Section 4 (b)
[82] https://www.dmr.nd.gov/oilgas/rules/rulebook.pdf
43-02-03-15. BOND AND TRANSFER OF WELLS.
[83] W. Va. Code, § 22-6-26; §22-6A-15
[84] http://www.pacode.com/secure/data/025/chapter78/chap78toc.html
§ 78.303. Form, terms and conditions of the bond.
[85] http://wogcc.state.wy.us/
[86] http://www.ndhealth.gov/aq/oilgaswell.aspx
[87] http://www.dep.pa.gov/Business/Air/BAQ/Permits/Pages/default.aspx#.VnsDJ5ODGko
[88] http://www.rrc.state.tx.us/about-us/resource-center/faqs/oil-gas-faqs/faq-flaring-regulation/

[89] OKLA. ADMIN. CODE § 165:10-3-15 (b).

[90] https://www.colorado.gov/pacific/sites/default/files/003_030614-729AM-R3-6-7-fact-sheet-003_1.pdf page4

[91] http://soswy.state.wy.us/Rules/RULES/7928.pdf Section 39. Authorization for Flaring and Venting of Gas (b)

[92] North Dakota Century Code 38-08-06.4

[93] OAC 1501:9-9-03 (k) 1509.20 ORC, 1501:9-9-05 (B) OAC

[94] http://www.dep.state.pa.us/dep/deputate/airwaste/aq/permits/gp/Comparison_Table_CSSD-Colorado-PA-Ohio-WV-EPA-Air_Standards-2014-05-28-1530.pdf http://apps.sos.nm.gov/adlaw/csr/readfile.aspx?DocId=1450&Format=PDF

[95] 1. http://www.dep.state.pa.us/dep/deputate/airwaste/aq/permits/gp/Comparison_Table_CSSD-Colorado-PA-Ohio-WV-EPA-Air_Standards-2014-05-28-1530.pdf
2. http://paenvironmentdaily.blogspot.com/2013/08/dep-finalizes-air-quality-permit.html

[96] http://www.emnrd.state.nm.us/OCD/documents/SearchablePDFofOCDTitle19Chapter15created3-2-2012.pdf 19.15.18.12 CASINGHEAD GAS

[97] http://www.owrb.ok.gov/supply/watuse/gwwateruse.php

[98] http://www.swc.nd.gov/Data/swcftp/webfiles/Fact%20Sheet.pdf

[99] http://www.ncsl.org/research/environment-and-natural-resources/state-water-withdrawal-regulations.aspx

[100] http://www.nmlegis.gov/sessions/03%20Regular/FinalVersions/HB0976AGS.pdf

[101] http://texaswater.tamu.edu/water-marketing/acquiring-groundwater-and-surface-water.html

[102] https://westernstateengineers.files.wordpress.com/2011/11/wickerfall2011.pdf page 6-9

[103] http://www.swc.nd.gov/Data/swcftp/webfiles/Fact%20Sheet.pdf NDCC 61-04-02: Permit for beneficial use of water required

[104] 1. ORC 1521.16
2. ORC 1501.33

[105] http://www.dep.wv.gov/WWE/wateruse/Pages/FracWaterReportingForm.aspx

[106] http://www.ncsl.org/research/environment-and-natural-resources/state-water-withdrawal-regulations.aspx

[107] N. M. S. A. § 72-5-13

[108] COLO . CODE REGS . § 404-1:205A (b)(2)(A)(viii)

[109] 1.Id.

[110] 2.NDCC, 61-03-23 Penalties6Civil NDCC, 61-04-30 Penalties http://www.ose.state.nm.us/WR/

[111] http://www.wogcc.state.wy.us/

[112] http://www.rrc.state.tx.us/oil-gas/applications-and-permits/environmental-permit-types-information/recycling/

[113] COLO . CODE REGS . § 404-1:907(d)

[114] North Dakota Administrative Code 43-02-03-19.2

[115] 2 from Stronger 1&3.from http://oilandgas.ohiodnr.gov/industry/underground-injection-control 4. from ORC1509.22 (B)(2)(a)

[116] 2 from www.strongerinc.org/wp-content/uploads/2015/04/Final-Report-of-Pennsylvania-State-Review-Approved-for-Publication.pdf

[117] 1. from 19.15.17.12

[118] http://www.waterworld.com/articles/wwi/print/volume-28/issue-5/regional-spotlight-us-caribbean/fracking-wastewater-management.html

[119] http://files.dep.state.pa.us/OilGas/BOGM/BOGMPortalFiles/PublicResources/DEP_Chapter_78_Webinar_030915.pdf Page 6 http://files.dep.state.pa.us/OilGas/BOGM/BOGMPortalFiles/PublicResources/CHAPTERS%2078%20AND%2078a.pdf chapter 78a.

[120] N.M. Admin. Code 19.15.17

[121] Tex. Admin. Code tit. 16, § 3.8(d)(4)(G)

[122] OKLA. ADMIN. CODE § 165:10-7-20 (b).

[123] 43-02-03-19.3

[124] 1. from ORC 1509.22 (c)(6)
2. from OAC 1501:9-3-08

[125] W. Va. Code Sr. R. § 35-3-14 http://www.dep.wv.gov/oil-and-gas/Impoundments/Documents/Impoundment%20Refernces/WV%20Code%2035-04.pdf

[126] http://files.dep.state.pa.us/PublicParticipation/Public%20Participation%20Center/PubPartCenterPortalFiles/Environmental%20Quality%20Board/2016/February%203/Fact%20Sheet%20for%20Final%20Ch%2078%20Regulation.pdf

[127] N.M. Admin. Code 19.15.17

[128] http://www.rrc.state.tx.us/oil-gas/applications-and-permits/environmental-permit-types-information/swr8-summary/Tex. Admin. Code tit. 16, § 3.8(d)(4)(A)(d)(4)(III)

[129] OKLA. ADMIN. CODE § 165:10-7-16 e.7.

[130] COLO . CODE REGS . § 404-1:902(e)

[131] http://soswy.state.wy.us/Rules/RULES/7928.pdf Section 7

[132] NDAC 43-02-03-19.3. Earthen pits and open receptacles.

[133] 1509.072 ORC

[134] http://www.legis.state.wv.us/legisdocs/code/22/WVC%2022%20%20-%20%206%20%20-%20%2030%20%20.htm

[135] N.M. Admin. Code 19.15.17

[136] http://texreg.sos.state.tx.us/public/readtac$ext.TacPage?sl=R&app=9&p_dir=&p_rloc=&p_tloc=&p_ploc=&pg=1&p_tac=&ti=16&pt=1&ch=3&rl=9 16 T.A.C. §3.9(6)(A)(vi). RRC website, FAQ

[137] https://stateimpact.npr.org/oklahoma/2015/02/11/mapped-traffic-light-wells-in-oklahomasearthquakecountry/

[138] http://earthquakes.ok.gov/what-we-are-doing/oklahoma-corporation-commission/. http://cogcc.state.co.us/documents/about/TF_Summaries/GovTaskForceSummary_Engineering%20UIC%20Wells.pdf page2

[139] WY Rules and Regulations OIL GEN Ch. 4 section5 (c)

[140] North Dakota Administrative Code 43-02-05-04

[141] http://www.dep.wv.gov/oil-and-gas/GI/Forms/Documents/UIC%20APPLICATION%20 PACKAGE%2006-25-2014.pdf
page 7
page 10
[142] http://triblive.com/business/headlines/9178658-74/state-pennsylvania-activity
[143] http://www.rrc.state.tx.us/about-us/resource-center/faqs/oil-gas-faqs/faq-injection-and-dis-posal-wells/
[144] http://earthquakes.ok.gov/what-we-are-doing/oklahoma-corporation-commission/
[145] https://cogcc.state.co.us/forms/instructions/form%207%20specs-2.html
[146] WY Rules and Regulations OIL GEN Ch. 4 section 10
[147] North Dakota Administrative Code 43-02-05-12
[148] http://www.dep.wv.gov/oil-and-gas/Documents/UIC%20Permitting%20-%20Gene%20 Smith.pdf
Page 14 W. Va. Code St. R. § 35-4-7
25 Pa. Code § 78.125
[149] 40 CFR 146.23 - Operating, monitoring, and reporting requirements.
[150] N.M. Admin. Code 19.15.7.28
[151] http://www.rrc.state.tx.us/about-us/resource-center/faqs/oil-gas-faqs/faq-h-5/
http://www.rrc.state.tx.us/oil-gas/publications-and-notices/manuals/injectiondisposal-well-manual/pressure-test-report-summary-of-testing-requirements/general-testing-requirements/
[152] 1.OKLA. ADMIN. CODE § 165:10-5-6 (b)(1).
2.OKLA. ADMIN. CODE § 165:10-5-6 (d)(1)(A).
OKLA. ADMIN. CODE § 165:10-5-6 (d)(1)(B).
3.OKLA. ADMIN. CODE § 165:10-5-6 (f)(3)(A)
[153] https://cogcc.state.co.us/documents/about/TF_Summaries/GovTaskForceSummary_ Engineering%20UIC%20Wells.pdf
[154] WY Rules and Regulations OIL GEN Ch. 4 s 5
[155] http://www.occeweb.com/News/2016/02-16-16WesternRegionalPlan.pdf
http://www.occeweb.com/News/2016/03-07-16ADVISORY-AOI,%20VOLUME%20 REDUCTION.pdf
[156] http://www.platts.com/latest-news/natural-gas/houston/colorado-orders-shutdown-of-injec-tion-well-near-21808756
[157] https://www.dmr.nd.gov/oilgas/undergroundfaq.asp#mr10
[158] http://triblive.com/business/headlines/9178658-74/state-pennsylvania-activity
[159] http://www.tulsaworld.com/news/capitol_report/oklahoma-supreme-court-clears-way-for-earthquake-lawsuits-against-energy/article_546e6b34-4bfb-5298-a245-502c632a00dd.html

[160] https://legiscan.com/CO/text/HB1310/2016
Page 2
[161] http://headwaterseconomics.org/wphw/wp-content/uploads/state-energy-policies-wy.pdf
page 1, 6
[162] http://headwaterseconomics.org/wphw/wp-content/uploads/state-energy-policies-wy.pdf
[163] http://headwaterseconomics.org/wphw/wp-content/uploads/state-energy-policies-nd.pdf
Page 1
[164] http://headwaterseconomics.org/wphw/wp-content/uploads/state-energy-policies-nm.pdf
Page 4-5
[165] Id.
[166] http://headwaterseconomics.org/wphw/wp-content/uploads/state-energy-policies-co.pdf
page2
[167] http://headwaterseconomics.org/wphw/wp-content/uploads/state-energy-policies-nd.pdf
Page 2
[168] Id. Page 2
[169] 16 Texas Administrative Code §3.29
http://texreg.sos.state.tx.us/public/readtac$ext.
TacPage?sl=R&app=9&p_dir=&p_rloc=&p_tloc=&p_ploc=1&p_tac=&ti=16&pt=1&ch=3&rl=29
[170] OAC Chapter 10 165:10-3-10(b)
[171] Rule 205a
[172] http://sosswy.state.wy.us/Rules/RULES/7928.pdf
Section 45. Well Stimulation (d)
[173] N.D. Admin. Code
§ 43-02-03-27.1
1.g.
[174] http://codes.ohio.gov/orc/1509.10v1
Ohio Rev. Code §1509.10.
[175] http://www.legis.state.pa.us/cfdocs/legis/LI/consCheck.cfm?txtType=HTM&ttl=58&div=0&c hpt=32&sctn=22&subsctn=0
58 Pa. Cons. Stat. § 3222 Well reporting requirements
[176] NMAC 19.15.16.19 (B)
[177] https://oilandgas.ohiodnr.gov/portals/oilgas/pdf/ogc-brochure_062013.pdf

reflects problems in transparency. Also, public involvement is not required in decision making on enforcement of regulations. With water regulations, the TCEQ has yet to impose a limit on the withdrawal of water for fracking. The allowance of unrestricted amounts of groundwater for fracking leads to a shortage of water for agriculture and ranching. A balance needs to be struck over water use for production purposes irrespective of their types.

There exist problems in the structure of the organization by allowing two elected members to serve as top officials. Since the oil and gas industry is allowed to contribute to the election campaigns of the agency's elected officials and there exists no state limit on the donation amount, elected officials tend to show bias toward this industry that helped him or her to get elected. Also, the state's various rules and regulations, like the Rule of Capture and predominance of mineral rights over surface rights in land ownership, show a bias toward the oil and gas industry. Even the fee charged by the RRC for the plugging of inactive wells is much smaller in comparison with the actual cost and that charged by other oil and gas producing states like Oklahoma, Colorado, Wyoming, and North Dakota. Further, the state has yet to make changes to existing regulations on wastewater storage in underground wells that are located close to underground faults. As a result, the problem of earthquakes persists even many years after the peak activity of fracking ends. Although new regulations require seismic study for the permitting of new wells, no amendments have been made in requirements for existing wells that lie close to the fault zones. Another indication of state bias lies in the passage of the state law that has reduced the amount of money a local government can receive as an award from the jury for pollution by the oil and gas industry. Such a state law conveys the message that the interests of the industry precede those of public health and welfare. Even the many lawsuits filed against the EPA by the state to lower federal enforcement standards on emissions bear testimony of support to the oil and gas industry in Texas.

Conclusion

In Texas, the outdated land and water laws and the inadequate regulations on oil and gas drilling have proven to be conducive for oil and gas drilling activities in the state including fracking at a later time. The long tradition of drilling has helped the oil and gas industry to prosper and consolidate its position in the state. The industry has developed a cozy relationship with the state's dominant regulatory agency, leading to its capture and further promotion of its drilling activities. The industry's lobbyists have repeatedly sent messages to state officials on the many economic benefits of fracking to ensure government's support and prevent the introduction of new legislations on fracking. Though such efforts have helped in the expansion of fracking in the state, the regulatory actions of the state's agencies have evoked much criticism and have been blamed for a bias toward the energy industry.

In scrutiny of the regulatory actions undertaken by the administrative agencies, some critics, a few environmental groups and those people negatively affected by fracking, have raised their voices against the state's lackadaisical attitude toward regulation. Unfortunately, their efforts have proved to be of little or no avail. Their voices have been muted by the government officials when individual oil and gas companies have been asked either to pay a small penalty or settle the problem at the local level instead of seeking a statewide solution to it. Such state attitude and the existing deficiencies in the regulatory framework calls for the need to seek a balance between private and public interests in the state through the introduction of legislations that would not only help to internalize the social costs of oil and natural gas production but also provide adequate protection to people living close to the drilling sites.

References

Barringer, F. (2007, October 18). New battle of logging vs. spotted owl looms in west. *New York Times*. Retrieved from www.nytimes.com/2007/10/18/us/18owl. html.

Behrens, B. P. (2011). Rule 37 exceptions and small mineral tracts in urban areas: an argument for incorporating compulsory pooling into special field rules in Texas. *Tex. Tech Law Review*, 44, 1053.

Brennan, R. & Peacock, B. (2010). Condemnation Compensation. Texas Public Policy Foundation, Austin.

Calhoun, B. (2017). Lawmakers push for a fracking ban in Florida. Retrieved from www.mypanhandle.com/news/lawmakers-push-for-fracking-ban-in-florida/644940834. 17.

Cheren, R. (2014). Fracking bans, taxation, and environmental policy. *Case Western Reserve Law Review*, 64(4), 1483–1517.

Davis, C. (2012a). The politics of "fracking": Regulating natural gas drilling practices in Colorado and Texas. *Review of Policy Research*, 29(2), 177–191.

Davis, D. (2012b). Fracking shale gas and nonprofit environmental organizations. A paper presented at the American Society for Public Administration in 2012 at Las Vegas, Nevada.

Davis, C. & Hoffer, K. (2012). Federalizing energy? Agenda change and the politics of fracking. *Policy Sciences*, 45(3), 221–241.

Dresser, M. (2017, February 3). Fracking ban bill introduced in Maryland Senate. *Baltimore Sun*. Retrieved from www.baltimoresun.com/news/maryland/politics/bs-md-fracking-ban-20170203-story.html.

Earthworks. (2016). Surface owner protection legislation. Retrieved from www.oil-andgaslawyerblog.com/2009/03/surface-damages-for-oil-and-ga.html.

Elazar, D. J. (1994). *The American mosaic: The impact of space, time, and culture on American politics*. Westview Press.

Eoh, Y. (2014). Yes, No, Maybe So: Uncertainty in Texas Groundwater Withdrawal for Hydraulic Fracturing. *Houston Law Review*, 52, 1227.

Esch, M. (2015). New York state formalizes ban on fracking, ending 7-yr review. Associated Press, June 29, 2015. Retrieved from http://infoweb.newsbank.com/resources/doc/nb/news/15646D6174972E38?p=AWNB.

Fambrough, J., (2009). Minerals, surface rights and royalty payments. Real Estate Center, Texas A&M University, Technical Report 840.

FracFocus. (2016). About us. Retrieved from https://fracfocus.org/welcome.

Frosch, D. (2016, May 2). Colorado high court rules local bans on fracking are illegal. *Wall Street Journal*. Retrieved from www.wsj.com/articles/colorado-high-court-rules-local-bans-on-fracking-are-illegal-1462208729.

Ghoshray, S. (2012). Charting the Future Trajectory for Fracking Regulation: From Environmental Democracy to Cooperative Federalism. *Thurgood Marshall Law Review, 38*, 199.

Gradijan, F. (2012). State regulations, litigation, and hydraulic fracturing. *Environmental & Energy Law & Policy Journal, 7*, 47–84.

Gram, D. (2012). Vermont becomes 1st state to ban gas fracking. *Bennington Banner*. Retrieved from www.benningtonbanner.com/stories/vermont-becomes-1st-state-to-ban-gas-lsquofracking,157632.

Handy, R. (2017, February 19). Thousands of oil rigs go unmonitored. *Houston Chronicle*, A1 & A12.

Heinkel-Wolfe, P. (2015). Ban on hydraulic fracturing repealed. *Denton Record Chronicle*. Retrieved from www.dentonrc.com/local-news/local-news-headlines/20150617-ban-on-hydraulic-fracturing-repealed.ece.

Kahneman, D. & Tversky, A. (1979). Prospect theory: An analysis of decision under risk. *Econometrica: Journal of the Econometric Society*, 263–291.

Kingdon, J. W. (1994). Agendas, ideas, and policy change. *New perspectives on American politics*, 215–229.

Kulander, C. S. (2013). Shale oil and gas state regulatory issues and trends. *Case Western Reserve Law Review, 63*, 1101–1140.

Kurth, T. E., Mazzone, M. J., & Mendoza, M. S. (2012). American Law and Jurisprudence on Fracing—2012. *Rocky Mountain Mineral Law Foundation Journal*, 58(4), 1–84.

Lindblom, C. E. (1982). The market as prison. *The Journal of Politics*, 44(02), 323–336.

Malewitz, J. (2016). Eagle Ford town's residents disgusted by waste site's approval. *Texas Tribune*. Retrieved from www.texastribune.org/2016/05/03/railroad-commission-vote-tiny-nordheim-loses-waste/.

McFarland, J. (2009). Surface damages for oil and gas activities in Texas. Oil and gas lawyer blog. Retrieved from www.oilandgaslawyerblog.com/2009/03/surface-damages-for-oil-and-ga.html.

Nikolewski, R. (2016). Feds give thumbs-up to fracking off California coast. *San Diego Tribune*. Retrieved from www.sandiegouniontribune.com/sdut-offshore-fracking-california-2016may27-story.html.

North Texans for Natural Gas. (2015). Fracking funds Texas schools: A North Texans for Natural Gas Special Report. Retrieved from http://d3n8a8pro7vhmx.cloudfront.net/themes/55dc9a8f2213933dc0000001/attachments/original/1448900562/Fracking_funds_report_v9.compressed.pdf.

Opensecrets. (2015). Oil and gas industry profile: Summary 2015. Retrieved from www.opensecrets.org/lobby/indusclient.php?id=E01&year=2015.

Peretz, P. (1986). The market for industry: where angels fear to tread? *Review of Policy Research*, 5(3), 624–633.

Price. A. (2015). Oil and coal execs to lead Texas regulatory agency. Retrieved from www.statesman.com/news/news/oil-and-coal-execs-to-lead-texas-energy-regulatory/npkTY/.

Rahm, D. (2011). Regulating hydraulic fracturing in shale gas plays: The case of Texas. *Energy Policy*, 39(5), 2974–2981.

Ritchie, A. (2016). Fracking in Louisiana: The Missing Process/Land Use Distinction in State Preemption and Opportunities for Local Participation. *Louisiana Law Review*, 76(3), 810–853.

Ritchie, A. (2014). On Local Fracking Bans. *Natural Resources Journal*, 54(2), 255–317.

Rule, T. A. (2012). Property Rights and Modern Energy. *George Mason Law Review*, 20, 803–836.

Railroad Commission. (2016). History of the Railroad Commission. Retrieved from www.rrc.state.tx.us/about-us/history/.

Railroad Commission. (2016). Interim guidance for statewide rule 98: Standards for management of hazardous oil and gas waste. Retrieved from www.rrc.state.tx.us/media/2815/entiremanual.pdf.

Richardson, V. (2015). Jerry Brown turns on liberal environmentalists, rejects California fracking ban. Retrieved from www.washingtontimes.com/news/2015/jun/2/jerry-brown-rejects-california-fracking-ban-risks-/.

Supreme Court of Colorado. (2016). Supreme Court Case Number 15SC667. Retrieved from www.courts.state.co.us/userfiles/file/Court_Probation/Supreme_Court/Opinions/2015/15SC667.pdf.

Tamest. (2016). Tamest Shale Task Force: Oil and Gas regulatory overview. Retrieved from http://tamest.org/wp-content/uploads/2017/04/lori-wrotenbery-presentation.pdf.

TCEQ. (2016a). History of the TCEQ and its predecessor agencies. Retrieved from www.tceq.state.tx.us/about/tceqhistory.html.

TCEQ. (2016b). Requirements for reclaimed water. Retrieved from www.tceq.texas.gov/assistance/water/reclaimed_water.html.

TCEQ. (2016c). Who regulates oil and gas activities in Texas. Retrieved from www.tceq.texas.gov/assets/public/assistance/sblga/oil-gas/statewide_oilgas_prog_info.pdf.

Tomlinson, C. (2016, July 3). Who gets the bill to plug oil wells? *The Houston Chronicle*, B1.

Warren, G. (2014). Pooling clauses and statues in Oil, Gas & Energy Resources Law 101, Chapter 5, State Bar of Texas, Austin, TX. Retrieved from http://scholarship.law.tamu.edu/facscholar/711.

Woods, G. (2010). Mineral rights vs. property rights in Texas. eHow. Retrieved from www.ehow.com/list_6156127_mineral-vs_-property-rights-texas.html.

5 Fracking and its Impacts on the Environment

Introduction

In the United States, fracking has helped to usher in an era of shale revolution with an abundant supply of natural gas and oil that was once difficult to retrieve from the sedimentary and coal bed methane rock formations. Indeed, it has helped the nation to make rapid strides toward its goal of energy independence amidst the glut in world supply of crude oil. Along with it a new fear has cropped up – a slowdown in the development of renewable energy sources like wind and solar. Whatever maybe the case, energy development from unconventional sources has had profound impacts on our lives. On a micro scale, it has helped individuals in the United States to save money on fuel costs for transportation and heating of homes which largely influences their decisions on where to live and work in the present century. At the macro level, the economic and political impacts have been equally far reaching. It has threatened the financial well-being of nations that once heavily relied on drilling for oil and gas for their economic prosperity. Politically, it has helped to reshape the distribution of global power. The countries that once used their mineral wealth of oil and gas as an instrument to exert their influence on other nations either through promises of supply of oil or threats to cut it off, now have to rethink their strategies. Amidst the furor that fracking has created both domestically and globally, we cannot afford to overlook its environmental footprints.

At the beginning of the use of modern fracking technique in states like Texas and Pennsylvania, knowledge of environmental risks was limited and the nature of environmental impacts was difficult to predict. As fracking gained both in scale and momentum over the last decade, the environmental impacts have become all the more obvious. They have been found to be associated with various phases of production ranging from drilling to the collection of oil and gas, wastewater management, transportation, and operation of equipment at field sites. Since fracking can have significant impacts on local landscapes and environments, it warrants attention from decision makers and scientists in the planning of well sites and calculation of distance from wells in assessment of environmental impacts (Meng & Ashby, 2014).

The environmental impacts have also been blamed for causing psychological and social stresses in communities closer to fracking sites. Psychological stress has manifested in individuals' perceptions of sudden negative changes in their quality of life, feelings of frustrations, and those of being manipulated by the oil and gas industry along with uncertainty about the future, while social stress has divided families and communities when family members and community residents have adopted opposing views on fracking (Fisher, Mayer, Vollet, Hill, & Haynes, 2018). Thus with environmental degradation becoming an externality that cannot be overlooked or ignored any longer, this chapter presents some of the environmental impacts of fracking while trying to answer the question, what are the environmental impacts of fracking?

Environmental Impacts of Fracking

From the many complaints of residents in communities that have been negatively impacted by fracking's externalities, several environmental threats can be identified. Air pollution and threats to climate change have emerged as one of the most serious problems despite the oil and gas industry's claim that the use of natural gas in the utility sector is actually helping to prevent global climate change. Additionally, there exists the threat of contamination of the underground drinking water supply despite existing states' regulations on well casings, and the probability of accidents also cannot be ruled out. Other equally serious environmental impacts that have made the public fearful include earthquakes, water shortages, damage to infrastructure, and spikes in traffic fatalities from the increase in truck traffic. Even the noise emanating from various field operations and the bright lights used at drilling sites disturb the quality of life of individuals and the habitats of local flora and fauna in that area.

With the state and federal governments paying inadequate attention to the environmental problems arising from fracking and community residents lacking the resources to fight the big oil and gas companies, various environmental advocacy groups at the national and state levels like the Sierra Club, Natural Resources Defense Council, FracDallas, Occupy Well Street, and others have stepped in to lend support to local residents. They are creating awareness of measures that can be used at local and state levels of government to regain the status quo in their quality of lives and, in some cases, the larger national environmental groups have even filed lawsuits with state and federal courts to bring about changes in regulations on fracking. Additionally, they have rallied for the greater use of alternative energy sources but the fight against the big oil and gas companies has not been an easy one.

See Figure 5.1A for the logical sequence of events that lead to the deterioration of the environment. In Figure 5.1B the specific environmental

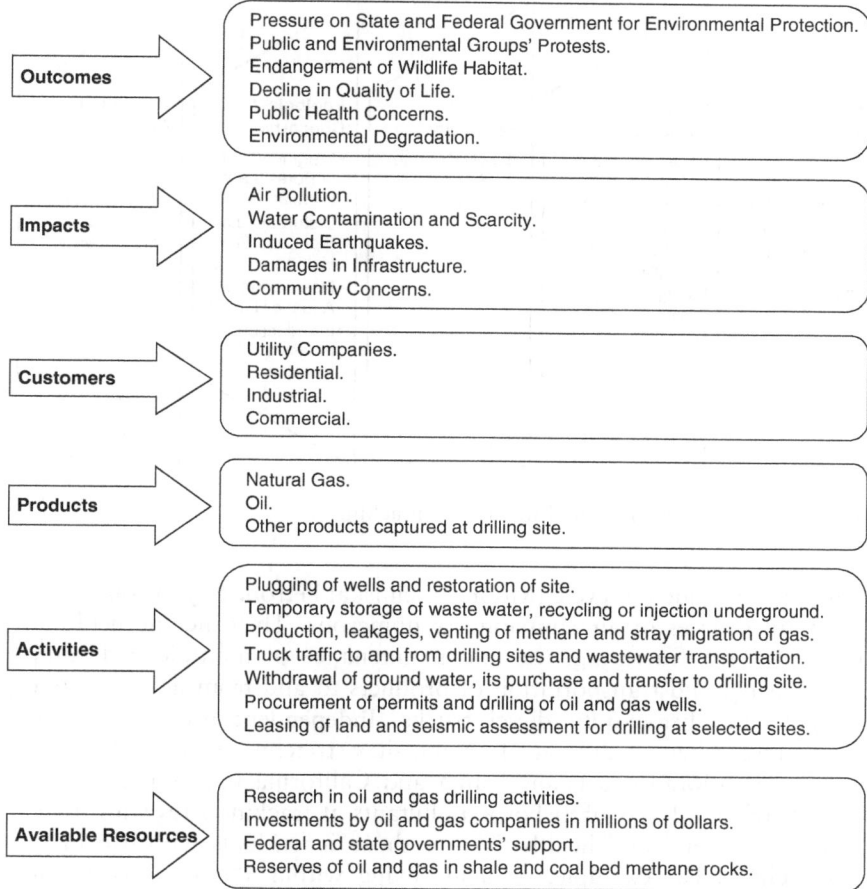

Figure 5.1A A Logical Model of Activities and Impacts of Fracking on Environment.

concerns have been laid out along with their sources. The various types of environmental pollution associated with fracking have been discussed in greater detail in the following sections.

Air Quality

Air pollution from a single well maybe low, but the cumulative impacts of air emissions from all the wells at a production site in a shale play pose a serious environmental problem. Even though the drilling process during fracking lasts from a few days to a few weeks depending upon the geologic structure of the region, air emissions continue over the life cycle of the

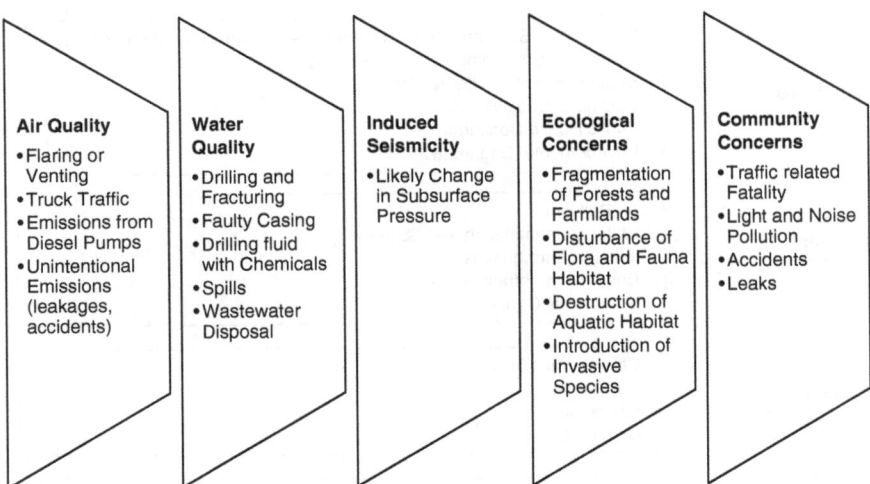

Figure 5.1B The Environmental Impacts of Fracking.

natural gas production cycle (Moore, Zielinska, Petron, & Jackson, 2014). The point sources of air emission are numerous. They include field sites during drilling, fracturing, use and maintenance of multiple pad equipment, flaring, and transportation of products to and from field sites both by light- and heavy-duty diesel trucks. Independent investigations by various researchers at different fracking sites stretching from the states of Texas and Colorado to Pennsylvania and California have confirmed the presence of high levels of airborne pollutants at fracking sites with potential environmental and health risks to those individuals who live in the close vicinity. For example, residents living within a range of 0.8 kilometers from a gas well have been found to be more susceptible to health risks from fracking than those beyond that distance (McKenzie et al., 2012). Since people attach more importance to public health concerns among children and adults, and to avoid further losses, many cities in the nation have issued either a moratorium or ban on fracking.

At any natural gas production site where fracking is used to release the trapped natural gas from sedimentary rocks and coal methane beds, air emissions usually occur in five phases. The air emissions from fracking sites are more likely to make one assume that it must be subject to control under the federal government's Clean Air Act. Unfortunately, that is not the case. As per the EPA's requirements, an emitting unit has to spew ten to 250 tons of air pollutants to be under the purview of the Clean Air Act. Since the sources of emissions at any fracking site, including the drilling rig, condensate tank, compressor, and other equipment, do not individually emit the required total amount of pollutants, they cannot be regulated under the existing federal law. If collective emissions from all units at a fracking site are taken into consideration, then the EPA's classification requirements are most likely to be met (Rundquist, 2014).

To determine the total amount of air emissions from a well site located on private land, it would require intricate and site-specific investigations which many states are not ready to undertake. To address this issue, the state of California has asked the local governments to identify hot spots in air emissions and take actions through their tools of planning and zoning. On the contrary, in Texas and surrounding states, the state governments have preempted the authority of local governments and they not only lack the regulatory authority but also the capacity or information to close the regulatory gaps (Rawlins, 2013). When it comes to fracking on federal leased land, the oil and gas companies are required by the Department of Interior's BLM to report their emissions. But the limited guidelines on measurements from BLM have made these companies use their own methods of measurements that vary from firm to firm (U.S. Department of Interior, 2001). It has also helped the smaller oil and gas companies to evade the task of reporting emissions.

In the discussion on emissions from natural gas and oil production sites, distinction has been made between methane and non-methane emissions in the following sections.

Methane Emissions

According to an EPA estimate, oil and natural gas drilling sites are the second largest stationary sources of greenhouse gases and accounted for 236 million metric tons of carbon dioxide equivalent in 2014 (GAO, 2016). Since 90 percent of natural gas is composed of methane and it is 20 times more potent than one pound of carbon dioxide as a greenhouse gas, its direct emission into the atmosphere partly helps to offset the gains made in the reduction of carbon dioxide emissions from the utility plants.

According to the EPA, the natural gas field sites in the U.S. account for one-third of the emissions of methane after landfill and industrial production sites. High levels of methane emissions have been detected from fracking sites in various parts of the U.S. using infrared cameras and other equipment. Examples of such locations include the Barnett Shale of Texas, Bakken Shale of North Dakota, Uintah Basin of Wyoming, and Marcellus Shale of Pennsylvania. Researchers from the Jet Propulsion Lab and California Institute of Technology have found that most of the methane hot spots in the four states' corners of Arizona, Colorado, New Mexico, and Utah were mostly from 250 sources that include gas wells, storage tanks, pipelines, and processing plants (Elliot, 2016). According to the U.S. Government Accountability Office (GAO) Report (U.S. GAO, 2010), 126 billion cubic feet of natural gas worth $58 million in federal royalty payments has been vented or flared from federal leased lands.

Table 5.1 Trends in Methane Emissions from Energy Sources

Methane (CH₄)	1990	2005	2010	2011	2012	2013	2014
Natural Gas Systems	203.8	177.3	166.2	170.1	172.6	175.6	176.1
Petroleum Systems	38.7	48.8	54.1	56.3	58.4	64.7	68.7
Coal Mining	96.5	64.1	82.3	71.2	66.5	64.6	67.6
Stationary Combustion	8.5	7.4	7.1	7.1	6.6	8.0	8.1
Abandoned Underground Coal Mines	7.2	6.6	6.6	6.4	6.2	6.2	6.3
Mobile Combustion	5.6	2.7	2.3	2.2	2.2	2.1	2.0
Incineration of Waste	≤0.05	≤0.05	≤0.05	≤0.05	≤0.05	≤0.05	≤0.05

Source: Figures obtained from EPA's Trends in Greenhouse Emissions, Table 2–4: Emissions from Energy.

Notes
Available at www3.epa.gov/climatechange/Downloads/ghgemissions/US-GHG-Inventory-2016-Chapter-2-Trends.pdf.
Please note all numbers are expressed in MMT CO2 Eq. (million metric tons carbon dioxide equivalent).

In the state of North Dakota, 30 percent of natural gas from the Bakken Shale once used to be flared in the absence of adequate natural gas gathering and transporting pipelines and processing plants (U.S. GAO, 2004). Over the years, with improvements in pipeline infrastructure, the amount flared has been reduced to 10 percent in 2016 (Energy Information Administration, 2016). In Texas, flaring is allowed during drilling, testing of wells, and repairs of well or compressors but it does require a permit (Table 5.2). Only 0.8 percent of the natural gas is flared in the state but venting and leakages still continue to take place. In the lack of consensus over the safe limit of venting of this gas from a compressor or any location

Table 5.2 Permits Issued in Texas for Flaring

Year	Number of Permits Issued for Flaring
2016	4,870
2015	5,689
2014	5,285
2013	3,012
2012	1,963
2011	651
2010	306
2009	158
2008	107

Source: Texas Railroad Commission.

Note
Available at www.rrc.state.tx.us/about-us/resource-center/faqs/oil-gas-faqs/faq-flaring-regulation/.

site (Percival, 2010a), methane emissions remain an environmental concern in all the states where fracking takes place.

The different methodologies used by states and the EPA in measurement of methane emissions have led to its underestimation in various parts of the nation (Brandt et al., 2014; Miller et al., 2013; Mitchell et al., 2015). In view of this problem and the fact that methane is a potent greenhouse gas, environmental groups have called for its reduction in emission from all oil and gas producing sites where fracking takes place. It prompted two environmental advocacy groups, the WildEarth Guardians and San Juan Citizens Alliance, to file a lawsuit against the EPA in 2009. In 2010, the groups reached an agreement with the EPA to reduce methane emissions through the introduction of a new rule and the revision of existing ones. The proposed new rule was introduced in 2011, followed by its signing and finalization in 2012. This new rule applied only to onshore natural gas production sites that were either built, modified, or refracked since August 2011. It required the use of a "green completion" process made mandatory from 2015 onwards. Green completion refers to the use of equipment to capture and separate mixed gases and liquids that escape from a completed natural gas well. The propane, butane, and liquefied natural gas captured through the green completion process have a market value and it takes approximately two years to reap the monetary benefits from gas recovery. Prior to the mandatory implementation period, oil and gas companies were encouraged to use the green completion process. Those companies that opted not to use the green completion process before the 2015 implementation deadline, either

Figure 5.2 Oil and Gas Producing Areas.

Source: U.S. Energy Information Administration (EIA). Available at www.eia.gov/todayin energy/detail.cfm?id=18691.

due to their lack of infrastructure or low-cost effectiveness, were asked either to burn or flare natural gas containing methane and other compounds from a completed gas well. According to the EPA, flaring is a much better option than releasing fugitive methane and volatile organic compounds directly into the atmosphere (Weinhold, 2012).

A year later in 2016, the Obama administration, in its commitment to take action on climate change and protect public health, once again targeted methane emissions from oil and natural gas production sites. In tandem with the President's Climate Action Plan, the EPA announced new rules to cut back on methane, volatile organic compounds, and other toxic emissions from both new and existing oil and gas wells including those not covered by the 2015 rule. The new rules aim at reducing methane emissions by 40 to 45 percent from 2012 levels by 2025. Based on 9,000 comments received on the 2015 methane reduction plan, the EPA removed the exemption clause on monitoring of low production wells and required the survey of compressors twice a year to detect and repair leakages. The new 2016 rule is expected to yield $690 million worth of climate benefits by 2025 and outweigh the estimated costs of implementation of $530 million in the long run. Additionally, the EPA introduced the Methane Challenge Program. It recognizes those oil and gas companies that have developed ambitious plans to reduce methane emissions. Information on best practices in methane emission reduction is collected by the EPA and then disseminated among other oil and gas companies to achieve a similar goal (Jones, 2016).

The oil and gas industry has opposed the new 2016 rule. It has argued that its big and small companies are already committed to emission reduction in their own economic interests. The new rules would simply add to their costs of production, especially at a time of stagnation and losses in the oil and gas industry. The state of West Virginia, the first to come to the aid of the oil and gas industry, has filed a lawsuit against the EPA to prevent further job losses in the industry. It has been supported by 12 other states including Alabama, Arizona, Louisiana, Kansas, Kentucky, Michigan, Montana, Ohio, Oklahoma, South Carolina, Wisconsin, and North Carolina (Geiling, 2016). North Dakota has filed a separate lawsuit on its own, and the state of Texas, which has already filed 12 lawsuits against the EPA since Obama took office in 2009, filed another one in 2016 in the Washington D.C. court. In a 2016 lawsuit, it challenged the federal government's authority to force oil and gas companies in Texas to detect and repair leaks that are invisible to the naked eye (Osborne, 2016b).

A year later in 2017, the EPA changed its stance on the reporting of methane emissions, much to the relief of oil and gas companies. After the nation's withdrawal from the Paris Climate Agreement under the Trump administration and in an effort to strengthen the partnership with the oil and gas producing states, this agency no longer require compliance with the Obama era rules. Realizing that the collection of information imposes an additional burden on oil and gas companies, the EPA has agreed to assess the necessity of collecting such information (Lavelle, 2017).

Non-Methane Emissions

The non-methane emissions from the natural gas fields include criteria pollutants of oxides of nitrogen and volatile organic compounds (the precursors of ground level ozone), particulate matter, and hazardous emissions of formaldehyde, benzene, toluene, ethylbenzene, and xylene. The origin of non-methane emissions can be partly traced to the light- and heavy-duty diesel trucks that bring in millions of gallons of water, sand, and chemicals to the production site and transport the end products from the field site. According to the Texas Department of Transportation's study of truck traffic in the Barnett Shale area of Fort Worth, "it takes 1,184 loaded trucks to bring one gas well into production, 353 loaded trucks to maintain it and 997 loaded trucks to refract the well after every five years." This truck traffic at a natural gas production site is roughly equivalent to eight million cars and an additional two million cars (Barton, 2012). The refineries that process shale oil also contribute to emissions of volatile organic compounds, such as observed in Los Angeles, California where refineries have been identified as the worst polluters. There they account for emissions of 1,600 to 3,200 tons of volatile organic compounds per year that contribute to local air pollution (May, 2012).

In Texas, air pollution is a serious problem in both urban and rural areas of the Barnett and Eagle Ford Shale plays, the two most active drilling sites in North Texas. Here, oil and gas companies with permits from the state government release tons of nitrogen oxides, carbon monoxides, sulfur dioxide, particulate matter, and hydrogen sulfide into the air. Additionally, high levels of benzene have been detected through the use of infrared video cameras by the TCEQ in the Barnett Shale area (Percival, 2010b). In the nine counties of the Dallas Fort Worth metropolitan area, the high level of ozone has been blamed for smog and respiratory problems among adults and children. Based on an air monitoring report, the EPA has labeled this area as an ozone non-attainment area. Here, in four out of the nine counties, the high levels of ozone have been traced to the release of volatile organic compounds from oil and gas operations. Even the TCEQ's state implementation plan has attributed the high level of ozone in the region to drilling for oil and gas in the four counties. Energy production in the identified counties contributes more volatile organic compounds than from all types of vehicular traffic combined in this area (Rawlins, 2013).

In urbanized areas of shale plays, residents living close to fracking sites have expressed environmental concerns at the deterioration in their quality of life. Though they may not be always aware of the identity of gases released into the air from the natural gas production sites, they certainly have experienced the adverse effects of them on their health. For instance, in the six counties of the Barnett Shale region of the Dallas Fort Worth metroplex area where fracking is most intense, there is high incidence of asthma among children, which has been considered above the state average (Tubb, 2010). Even occurrences of invasive breast cancer are also high in this region. In Flower Mound, a suburb of Dallas, residents have called for a state investigation into the high rate of children's leukemia in this area (Rawlins, 2013).

The deleterious effects of air pollution on public health in communities close to fracking sites have made the local residents upset and concerned. It has been partly responsible for their negative perception of oil- and gas-related socio environmental issues. To voice their complaints, residents have not only attended neighborhood and town hall meetings to oppose fracking activities but have also contacted their local and state government officials (Theodori, 2012).

At the local government level, local residents' complaints and requests for greater environmental protection have been only met with lukewarm responses. Since the passage of the Denton fracking ban, the state has imposed restrictions on local governments' authority to control oil- and gas-related operations, including fracking within their jurisdictions. As a result, ordinances are the only instrument that a local government can use to restrict the activities of the oil and gas companies. Under pressure from residents, local governments have either amended the old or introduced new ones. The ordinances that the oil and gas companies have failed to influence or resist at the local government level have not always proven to be effective. For example, the operators have invoked the grandfather clause to seek a reprieve from the new restrictions on their activities in the case of Denton, Texas (Fry, Briggle, & Kincaid, 2015)

At the state level, deterioration in air quality has prompted its monitoring. In Texas, there are 82 monitoring sites equipped to detect 146 types of air toxics (Rawlins, 2013). The monitors are considered important in the data collection process and in determining the type and level of air pollution in the shale plays, but there exists a poor correlation between air monitoring networks and oil and gas production sites (Robinson, 2012). The dispersed nature of natural gas drilling sites and the location of monitors at the periphery of the shale plays have stood in the way of effective air pollution control (Song, Morris, & Hasemyer, 2014). Also, the installation of air quality monitors by the EPA and state agencies (especially for ozone emissions) mostly in the urban areas has failed to capture the seriousness of the problem in rural areas. Consequently, there exists little or no baseline and routine air quality measurements in the industrialized rural and suburban communities where fracking has left its footprints (Moore et al., 2014).

When it comes to the evaluation of collected data, the state of Texas does not use clear-cut standard values but instead Air Monitoring Comparison Values (AMCV). The AMCVs have been established with the help of guidance documents and not through any rule making procedure. In interpretation of air data, monitoring values below the AMCV are regarded as non-threatening while those above it call for in-depth investigations. Also, the biases of politicians and state officials related to their economic and political beliefs can interfere in the interpretation of monitoring values. Often these biases influence the level of importance attached to the problem and therefore the level of environmental protection offered to those individuals living near the fracking sites (Rawlins, 2013).

There also exists the problem of the escape of fugitive hazardous gases from vents and leakages at gas well production sites. Therefore, to address the problem, the EPA has developed a program with a practical solution to the problem, called the Gas STAR, which offers an online directory of service providers who can assist the oil and gas companies in identifying emission sources, quantify the amount of emissions, and make recommendations for the capture of those gases that have a market value. Texas offers a similar program through the TCEQ that offers a small business assistance program and a site visit plus (SAV+) program. From the free site visits of the SAV+ program, a facility owner can receive assistance in identification of performance issues, pollution prevention tips, and information on innovation strategies (Percival, 2010b). With federal and state governmental help available, the onus lies on the oil and gas companies to seek help from outside sources in their mitigation efforts. To what extent such free advisory services are used by oil and gas companies calls for the evaluation of programs and examination of governmental outreach efforts to determine their effectiveness in air pollution control.

Other Air Emissions

Besides the emission of harmful gases, another air pollutant that poses an occupational health hazard is respirable crystalline silica, a major constituent of sand, clay, and stone materials. Silica sand is used in fracking as a proppant, that is, to open up fractures in the dense shale rocks and widen the pathways for the escape of oil and gas from the rock pores (Wethe, 2016). The National Institute for Occupational Safety and Health (NIOSH), recognizing the hazards of silica, has conducted field studies in oil and gas fields that have confirmed workers' exposure to silica dust through inhalation during the mining phase, transportation, and in preparation of the fracking fluid in the oil and gas industry. A NIOSH study of 11 fracking sites in five states (Arkansas, Colorado, North Dakota, Pennsylvania, and Texas) from 2009 to 2011 has revealed that workers have been exposed to high levels of respirable silica in 79 percent of the 116 sampled cases. Their exposure to silica was higher than NIOSH's permissible limit of 0.05 milligrams per cubic meter and in 31 percent of the sampled cases, exposure to silica was ten times higher than the real exposure limit (NIOSH, 2012). Usually, individuals' prolonged exposure to silica dust at work sites enhances their risks of developing silicosis, a lung disorder that impairs the functions of lungs and causes respiratory and other disorders (Rundquist, 2014; Nall, 2013).

In response to growing concerns of workers' exposure to silica in the oil and gas and other industries, the Occupational Safety and Health Administration (OSHA) issued its final rule in June 2016. As per the new rule, the permissible exposure limit has been set at 50 micrograms per cubic meter of air averaged over an eight-hours shift. Additionally, it requires the industry to use engineering control to reduce workers' exposure to silica, provide medical exams to workers with high risk of exposure, and aids small businesses in enforcing the standards (OSHA, 2016). Another occupational hazard is the risk of exposure to radiation at drilling sites. Workers drilling into rocks that contain naturally radioactive materials like radon, thorium, and uranium can easily get exposed to high levels of radiation (McDermott-Levy, Kaktins, & Sattler, 2013).

The various pathways of air emissions have aroused the interests of scientists, non-profit organizations, and the private industry. Many times, when private companies' scientific findings tend to conflict with the interest of the industry, they are likely to be suppressed, as seen in the case of Exxon. The company's scientists were aware 11 years ago that burning fossil fuels would bring about a climate change but the company refused to acknowledge it and spread misinformation. Such information became available upon investigation by the Massachusetts attorney general (AG) and bears a close resemblance to what had happened in the tobacco industry in the past, raising the question of ethics in business practices. States in support of the oil and gas industry have condemned the Massachusetts AG's investigation on the grounds that it is not the job of the government to investigate what the company knew about climate change and when it knew. Further, in support of the oil and gas giant, which has filed for an injunction to block the demand for a decade's worth of oil and gas documents, the state of Texas and ten other states have filed an amicus brief protesting at such investigative actions (Zelinksi, 2016). Once again, this brings to the forefront states' bias toward the oil and gas industry and their tendencies to overlook the environmental impacts of oil and gas development.

Water Quality

Drilling for oil and gas in shale formations requires a large quantity of water. In examination of water use for fracking, there exists much ambiguity. The actual amount of water required for each well usually depends upon the geology of the shale, the number of fracturing stages, and the amount of flow back water generated. As a result, it can vary from two to eight million gallons of water in the sedimentary rock formations. The operators prefer to use groundwater for two reasons. First, groundwater is much less regulated than surface water (Freyman, 2014). Second, it has fewer impurities and does not impact the efficiency of fracking fluid (Eoh, 2014).

The large amount of water required and the types of drilling procedures used have made many critics express concern over the eligibility of groundwater use in fracking. This is because at the time of the writing of the state's water code, vertical drilling was the common procedure and horizontal drilling was unknown. To restrict water use for fracking purposes, some water districts have gradually started imposing their own regulations to control the use of city water. For example, the Evergreen Underground Water Conservation District, stretching across the four counties of Karnes, Frio, Wilson, and Atascosa in the Eagle Ford, added fracking in 2008 to its list of activities with preexisting limits on water use (Kurth, Mazzone, & Mendoza, 2012; Galbraith, 2013; Wittmeyer, 2013). Other underground water conservation districts like the Hemphill County and High Plains are also developing their own restrictions on water use including fracking. Also, in recent years, some municipalities, like Grand Prairie, Texas, have started imposing their own restrictions on water use for fracking that was unheard of in the past. In fact, the city of Arlington does not allow the use of groundwater withdrawn from wells within the city for use outside its boundary limits (Kulander, 2013).

Since fracking is done in some of the high-water stress or drought areas, its diversion to oil and gas production activities poses a serious threat to the local economy. For example, in Texas, 28 percent of wells in the Eagle Ford area and 87 percent of wells in the Permian Basin are located in areas of high to extreme water stress (Freyman, 2014). According to the TCEQ, there are 30 communities in the state that are at a serious risk of water shortages, many of which are in the shale plays. Faced with years of droughts, water overuse, and large withdrawals of water for fracking, they are finding their water supply being gradually depleted in reservoirs and underground aquifers (Goldenberg, 2013). According to a United States Geological Survey (USGS) report, the groundwater depletion rate has increased since 1950. The maximum increase took place during 2000 to 2008, with an average depletion rate of $25\,km^3$ per year compared to an overall average rate of $9.2\,km^3$ during the period of 1900 to 2008 (Konikow, 2015). See Figure 5.3 below.

Another water problem is its contamination when additives like sand and chemicals (disclosed and proprietary) are mixed to form the fracking fluid. This fluid is then injected under heavy pressure to fracture the rocks and stimulate the oil and gas flows from the unconventional wells. Some of the water used in fracking finds its way back to the surface as *flowback* water. The flowback water mixes with the natural water in the geologic rock formations, including radioactive substances, which leads to a further deterioration in its quality. Then there is the *produced* water that flows back to the surface along with the oil and natural gas during the production process. The total amount of wastewater generated in fracking is

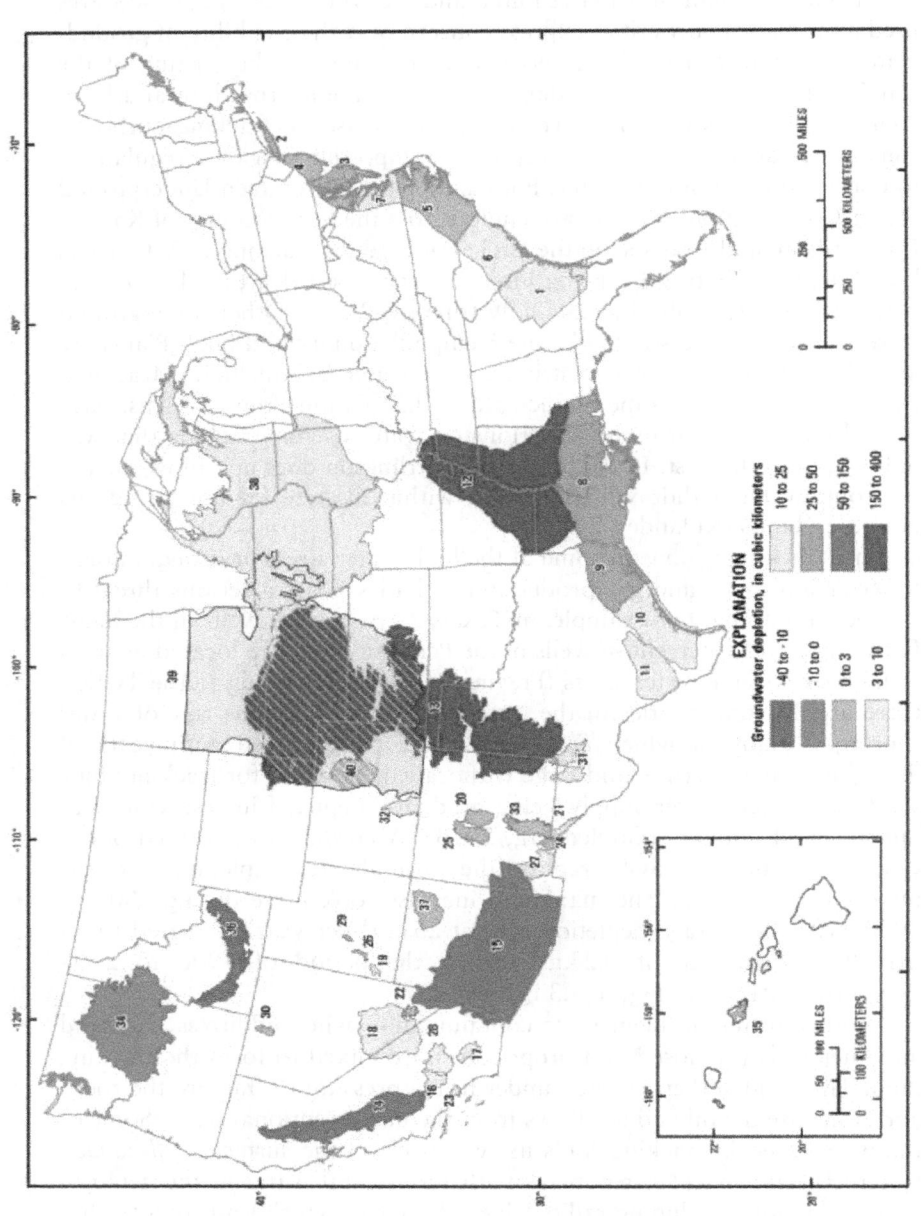

Figure 5.3 The Depletion of Groundwater in U.S.A.

Source: USGS. Groundwater depletion in the United States (Konikow, 2015).

Table 5.3 Water Consumption and Wastewater Production in Selected States Since 2005

States	Number of Wells Fracked	Water Consumed in Fracking (million gallons)	Wastewater Produced in 2014 (million gallons)
California	3,405	237	1,057
Colorado	22,615	19,142	3,139
North Dakota	8,224	14,891	unavailable
Ohio	1,594	7,771	313
Oklahoma	7,421	19,582	unavailable
Pennsylvania	9,223	24,732	1,821
Texas	54,958	120,215	unavailable
West Virginia	2,670	7,651	unavailable

Source: Environment Texas Research and Policy Group, Fracking by the Numbers (2016).

either stored on the surface or underground. Recycling of wastewater is an option, but only 3 percent of water used in fracking in the Barnett Shale is recycled because of the cost dynamics. It is much cheaper to inject wastewater underground (Cook, Huber & Webber, 2015). Other challenges in wastewater management include its runoff to neighboring lands mixed with storm water or from accidental spills. Federal regulation on wastewater management from fracking is very limited. Even though states play a major role in regulating water for fracking, deficiencies in state regulations pose a serious threat to the aquatic environment.

Drinking Water Contamination

The contamination of drinking water wells in communities located close to fracking sites has only helped to add to the resentment against fracking. The oil and gas industry has repeatedly denied that fracking leads to water contamination, and scientific studies have not yet been able to establish a direct link between fracking and underground water contamination. Nevertheless, investigations led by the EPA and other state agencies in various parts of the nation have found oil and gas drilling activities to be responsible for the contamination of drinking water supplies in private wells by indirect means.

Natural gas is located deep underground and its extraction calls for drilling through layers of rock that contain drinking water supply for many local communities. When a well is not properly cased or sealed, natural gas is likely to escape and contaminate sources of underground drinking water, irrespective of the stringency of state regulations. Even corrosion of old and abandoned wells serve as migration pathways and are responsible for contamination. Other likely pathways for the escape of oil and gas include underground fractures formed during the fracking process (EPA, 2011).

These fractures facilitate the migration of oil or natural gas to the layer of rock containing the drinking water supply. This is a serious concern in coal bed methane areas where natural gas not only tends to lie at shallow depths but also lies closer to the underground drinking water sources (Cooley & Donnelly, 2012). Even short-term changes in water quality like odor, color, and turbidity are brought about by vibration and changes in pressure during the drilling of wells (Groat & Grimshaw, 2012).

With the widening of the scope of contamination of underground drinking water supplies as result of fracking, some private well owners have experienced deterioration in their drinking water quality in various parts of the country. For example, in Pavilion, Wyoming, local well owners in 2008 complained of changes in odor and taste in their drinking water. In Dimmock, Pennsylvania, a large well explosion in 2009 shocked the local community residents. Images of faucet water catching fire in the documentary film Gasland (Fox, 2010) because of oil and gas operations have helped to create awareness and capture the attention of millions of people nationwide. It has also helped to put pressure on state and federal agencies for investigation and remediation of the problem.

With methane making up 90 percent of natural gas and it being identified as the main culprit in deterioration of drinking water quality, many people have been asking why natural gas is not regulated by the state or EPA under the National Primary Drinking Water Regulations. The answer is simple. Since methane does not affect the taste, color, and odor of drinking water, it is not subject to regulation. In high concentrations methane can cause explosions, and its inhalation can lead to asphyxiation (Jackson, Pearson, Osborn, Warner, & Vengosh, 2011). To minimize such hazards, the U.S. Department of Interior (2001) has called for its venting at well heads when present in water at concentrations above 28 milligrams per liter and issuance of warnings among local residents when detected in drinking water wells above ten milligrams per liter (Vidic, Brantley, Vandenbossche, Yoxtheimer, & Abad, 2013).

Precluding methane, contamination of groundwater can also result from the escape of other chemical compounds present in the fracking fluid. In Texas, a study of 550 groundwater samples drawn from private and public water supply wells in the Barnett Shale play has revealed high levels of ten different metals and 19 types of chemical compounds including benzene, toluene, xylene, and ethylbenzene (Hildenbrand et al., 2015). In another study of private drinking water wells located 3 km from active gas wells in the Barnett Shale area, analyses of 100 samples have revealed arsenic, selenium, strontium, and total dissolved solids in quantities that exceeded the EPA's Maximum Contaminant Limit (MCL). Methanol and ethanol were also found in 29 percent of the cases along with other chemicals above the MCL in wells randomly that were dispersed in the shale play. The source of their origin was not traced but comparison of historic values with those of current ones revealed a significant increase in amount suggesting a higher risk of contamination of private wells when located closer to drilling sites. Since most of the chemicals were associated with active fracking in the region, the findings have called for the monitoring and analysis of groundwater quality on a regular basis here (Fontenot, 2013).

There exists the additional risk of bacterial contaminations at fracking sites. It can occur in drilling fluid, which includes mud and water, proppants, and storage tanks. To prevent the growth and proliferation of bacteria both on the surface and underground, different types of biocides are used. Some of these biocides are biodegradable while others are not and can become toxic under certain conditions. When such toxic biocides escape and reach the groundwater, they cause contamination (Kahrilas, Blotevogel, Stewart, & Borch, 2016).

Surface water contamination is another hazard arising from fracking. The natural or flowback water, along with the produced water, flowing out from underground wells contains dissolved salts, metals, chemicals, and petroleum hydrocarbons in varying amounts. The variable quality of these two types of water, which constitute about 15 to 85 percent of the water used in fracking, poses a challenge in its disposal and management to both producers and regulators.

Management of Wastewater

Some of the options in the management of wastewater range from its injection into deep wells to discharge into surface water bodies, treatment in commercial or municipal water treatment, and reuse in fracking with or without treatment. Whatever maybe the final choice, the decision usually rests on economics, EPA and state regulations under the National Pollutant Discharge Elimination System, amount of water generated, and availability of facilities for storage or treatment. For example, in the state of New Mexico and Marcellus Shale areas of the north eastern U.S., the wastewater is reused in fracking due to water shortages in the region and the low cost of such water.

In the states of Texas and Oklahoma, most of the wastewater is injected deep underground into Class II injection wells, mainly used for storage of non-hazardous waste. In those cases, where the wastewater from fracking is stored in storage pits, embankments, or tanks at the well site prior to transportation offsite into Class II injection wells, the risk of contamination of surface water through runoffs and spills cannot be eliminated. There has been accidental spillage or release of wastewater temporarily stored in the "frac ponds." For instance, brown and foamy water stored for fracking in a lined pond in the Eagle Ford Shale play of Texas was accidentally released into two neighbors' properties and the surface road. The incident led to a regulatory dilemma involving two regulatory agencies of the state as to who had the jurisdiction over it. Ultimately, the Rail Road Commission claimed responsibility and worked with the local operator to

Table 5.4 Management of Wastewater in Shale Formations

Shale Formations	Water demand per well	Type of disposal method mainly used	Reuse and its potential
Marcellus	80,000–85,000 gal for drilling; 3.3–5.5 Mgal for fracturing.	Disposal well injection, a small portion to wastewater treatment plant.	Over 90% of flowback and produced water are reused, most of which are stored for future hydraulic fracturing.
Barnett	250,000 gal for drilling; 3.8 Mgal for fracturing.	Disposal well injection for most flowback and produced water.	No reuse, but have a high potential because of new regulations and potential water shortage due to droughts. Texas Railroad Commission encourages reuse through recycling.
Fayetteville	60,000 to 65,000 gal for drilling; 2.9 to 4.9 Mg for fracturing.	Disposal well injection for most flowback and produced water.	No reuse, but have a high potential because of new regulations and potential water shortage due to droughts.
Haynesville	600,000–1,000,000 gal for drilling; 5 Mgal for fracturing.	Disposal well injection for most flowback and produced water.	No reuse. Poor quality makes it less attractive. Drilling mud is more feasible for reuse for its relative high quality.
Eagle Ford	125,000 gal for drilling; 2–13.7 Mgal for fracturing.	Disposal well injection for most flowback and produced water.	No reuse, but have a high potential because of new regulations and potential water shortage due to droughts. Texas Railroad Commission encourages reuse through recycling.

Source: Information obtained from Table 2, Summary of produced water management in major shale plays in Review of Flowback and Produced Water Management, Treatment and Beneficial Use for Major Shale Gas Development Basins by Ma et al. (2014).

Note
Mgal refers to million gallons.

ensure the safety of local owners and to make sure that such an incident will not be repeated (Hiller, 2016).

Additionally, the location of injection wells far away from densely populated areas due to legal restrictions has led to their siting in rural areas. This has given rise to the problem of environmental injustice (Bullard, 2000). For instance, in southern Texas in the Eagle Ford area, disposal wells have been found to be disproportionately permitted in areas with a high minority population which is living in poverty (Johnston, Werder, & Sebastian, 2016). Another problem is wastewater from injection wells migrating to surface and freshwater aquifers through cracks, leaks, or abandoned wells. For example, in south eastern Texas, the groundwater close to wastewater wells has been found to have a higher concentration of chloride and bromide than groundwater farther away. Such a finding helps to question the validity of the exemption of hazardous waste from fracking under the RCRA and why states do not prevent it in the first place rather than penalizing the polluter after the incidence. Further, the state's regulatory agency (RRC) has failed to keep its promise made in 1982 to track if oil and gas companies have injected toxic waste into zones that hold underground drinking water. Even the EPA has no records for granting such exemptions in Texas, and it has asked the state to prioritize the gathering of data (Malewitz, 2016).

The media has helped to raise public awareness on various issues related to wastewater and its management. Local environmental groups have also stepped in to help people identify problems and navigate the bureaucracy in search of remedial actions. With the identification of toxic elements in wastewater from fracking and states' lackadaisical attitude toward regulation on wastewater injection, a coalition of national environmental groups composed of the Natural Resources Defense Council, Environmental Integrity Projects, and six others filed a lawsuit against the EPA in the District of Columbia court in May 2016 to seek redress. Under the RCRA, the EPA is expected to review regulations and state guidelines every three years. Unfortunately, the EPA has failed to do that since 1988. Hence, the recent lawsuit has asked the EPA to review and update regulations on oil and gas waste and requested the court to set the deadline for updates (Fragoso, 2016). Also, in view of the enormity of the problem, more regulatory actions definitely need to be taken by states and their water development districts or boards to address the pressing problems in water use and disposal of wastewater.

Induced Seismicity

In the last decade, a spate of earthquakes in regions where such activities have been rare or uncommon have only aroused the suspicions of local people who have blamed fracking. These accusations have been denied and ignored by the oil and gas industry in the absence of evidence linking earthquakes to fracking. In review of regulations at the state level to address the problem, it is evident that regulations on such an unforeseen activity from fracking were absent at the very beginning of fracking.

In recent years, scientific studies have been conducted to investigate the cause of earthquakes in those states where fracking is common and earthquakes have been absent or rare in the past. Review of past studies and court and other available documents have helped to reveal the occurrence of small earthquakes probably induced by extraction of oil and water as far back as 1925 along the Gulf coast (Hirji, 2016). In 2016, the U.S. Geological Survey (USGS) for the first time issued a map showing potential earthquakes from natural and human induced activities. See Figure 5.4 below.

The state most affected by earthquakes since fracking started has been Oklahoma. Here, earth tremors ranging in magnitude from 2.2 to 3.3 on the Richter scale have rattled homes and caused small damages to buildings. News of such earthquakes made headlines in local, state, and national newspapers and drew the attention of people nationwide. Other states that experienced similar quakes close to fracking sites include Texas, Kentucky, Ohio, and Colorado. In Texas, prior to 2008, earthquakes of 3.0 magnitude occurred on average only twice per year. After 2008, their frequency

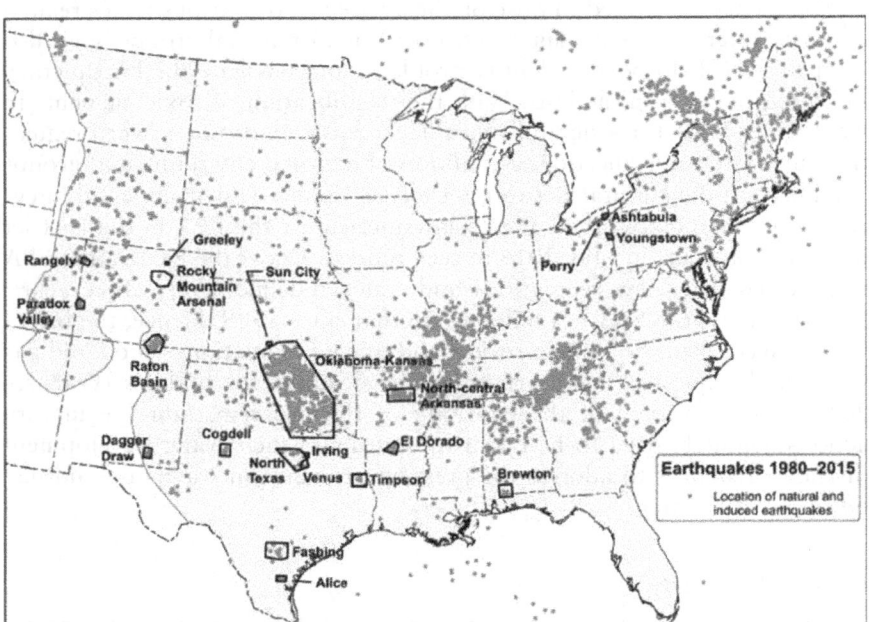

Figure 5.4 Earthquakes Since 1980 and Recent Areas Impacted by Induced Seismicity.

Source: USGS, 2016 www.usgs.gov/news/induced-earthquakes-raise-chances-damaging-shaking-2016.

increased to 12 per year coinciding with the increase in fracking activities in North, South and East Texas (Hirji, 2016). During the period from 2014 to 2015, a spate of earthquakes in North Texas drew the attention of researchers at the University of Texas (UT), Austin and Southern Methodist University in Dallas. They launched their studies on earthquakes in the Dallas Fort Worth area, which is densely populated with an urban population of over four million people and no prior historic records of earthquakes since 1850. Another compelling reason was its location in the Barnett Shale play, which had experienced a sudden increase in drilling and fracking for natural gas since 2002 (Frohlich, Haywood, Stump, & Potter, 2011).

Although the studies have found no direct evidence of earthquakes caused by drilling, fracking, or extraction of natural gas from the underground wells, analysis of tremors has suggested otherwise. For example, in the small city of Azle in the Dallas Fort Worth area and located at close proximity to an underground injection well, the occurrences of several earthquakes have suggested that wastewater injection and removal of flowback or brinewater during drilling were most likely responsible for significant changes in subsurface stress and subsequent quakes in the area (Frohlich et al., 2011; Hornbach et al., 2015).

The scientific findings have evoked different responses from governmental agencies. At the federal level, they have prompted the U.S. Geological Survey (USGS) scientists to issue warnings to the oil rich areas of Oklahoma City and Dallas Fort Worth. Here, the risk of earthquakes is on a par with that of California, a state prone to stray incidences of earthquakes. According to the USGS's estimate, the states that are at most risk from human induced earthquakes include Oklahoma, Kansas, Texas, Colorado, New Mexico, and Arkansas (Osborne, 2016d).

At the state level, responses have been slow. In Oklahoma, regulators have ordered a cutback on pumping of wastewater into injection wells by 40 percent in the central part of the state. In Ohio, an unofficial moratorium on permits for injection of wastewater underground was issued for 11 months. Currently permits are now issued but at a slower rate. Regulators are also requiring more thorough investigation of the geology of a region prior to injection which is making the process of wastewater injection more expensive than before (Schmidt, 2013).

In Texas, the regulators initially disputed the connection between fracking and earthquakes and claimed the studies' findings to be subjective in nature. In the absence of substantial evidence, the Texas RRC has decided not to put the blame yet on oil and gas operations. It has cleared two energy companies in North Texas from local residents' accusations of causing two dozen minor earthquakes in the cities of Reno and Azle. Unlike Oklahoma, which has hired qualified seismologists to study the problem, Texas has hired a rancher to review the problem (Tomlinson, 2016). The oil companies have lauded the regulatory agency for its actions but public outcry and pressure from environmental groups have made the Texas Railroad Commission (TRRC) adopt a new rule on permit applications for disposal wells in 2016. Already lagging behind the states of Oklahoma, Ohio, and Arkansas in the adoption of measures to control human induced seismicity, the state has called for the collection of data on historic earthquakes in a 100-mile radius of the proposed site (Malewitz, 2015). Further, the state's lawmakers, urged by scientists to monitor earthquakes and after long deliberations, finally approved about $4.5 million in 2015 for TexNet to enhance the monitoring network of seismic monitors used in subsurface data collection (Hiller, 2016).

The continuation of earthquakes closer to the injection wells, such as the 5.6 magnitude earthquake in September 2016 in Oklahoma's energy producing corridor, that also rattled the mid-western states and the southwest of Texas (Miller, 2016), raises questions about states' initiatives to stop such tremors. With no regulation on earthquakes, Oklahoma has suspended all operations in 37 wastewater wells while Texas is yet to take such actions on existing wastewater wells. The oil and gas friendly states need to reevaluate their stance on environmental protection to citizens and review the existing measures on management of wastewater from oil and gas drilling operations. Without succumbing to the pressure from the oil and gas lobbying groups, they need to amend the existing regulations and or introduce new ones to reduce human induced seismicity instead of denying the existence of a link between wastewater injection activities and earthquakes. States need to pay attention to scientific evidence and exert more control on injection of wastewater into disposal wells. By making regulations more stringent and increasing the cost of non-compliance, the states can promote greater reuse of the water through the recycling process. Since recycling of the wastewater calls for investment by oil and gas companies into expensive recycling facilities that are suited for treatment of such waste, the states can offer incentives to undertake more such activities.

At the federal level, the time has come for the EPA to review the exempt status it has granted to the oil and gas industry in management of wastewater. On the basis of current evidence offered by scientific studies, the exempt status ought to be removed, and the EPA should consider the treatment of wastewater from fracking as hazardous waste for injection purposes into deep underground wells, whether or not it is tainted by diesel. Since accidents can occur at any stage in the management of wastewater including transportation and storage, it is equally important for government agencies to know the identity of all toxic chemicals used in fracking fluid as they are a large component of the wastewater.

Ecological Impacts

The advancements in fracking technology have helped to efficiently reap oil and gas from unconventional sources but the expansion of well pads over a larger area has meant increased threats to wildlife and ecosystem protection. In traditional drilling, two to three acres of land need to be cleared – fracking requires four to five acres of cleared land. This unconventional method of drilling needs more space not only for drilling of multiple wells but also for storage of fluids, chemicals, drill cuttings, and equipment. The larger the size of cleared land, the more susceptible it becomes to environmental threats arising from runoff and siltation. Additionally, fracking leaves behind visual scars and causes fragmentation of agricultural and forested lands. If there are endangered or protected species living in those areas and their geographical range is limited, like the sage grouse and the prairie chicken, then the problem becomes all the more acute.

In some fracking sites, the withdrawal of surface water from streams and rivers can prove to be catastrophic to many aquatic species that require a certain level of water to thrive (Robbins, 2013). The bright lights and sound from fracking can disturb the feeding, breeding, and rest patterns of flora and fauna and degrade the local ecosystem (Burton et al., 2014). A study of animal owners living close to fracking sites in Colorado, Texas, Louisiana, Pennsylvania, New York, and Ohio has shown that animals like cows, horses, goats, chicken, llamas, dogs, cats, and fish in ponds are highly susceptible to environmental threats and deaths from exposure to contaminated springs and well water, dumping of wastewater into creeks and land, flaring, pipeline leaks, and compressor station malfunctioning. In most states, in the absence of adequate testing and disclosure of chemicals used in fracking fluid and the signing of non-disclosure agreements between oil and gas companies and owners of cattle and other animals, it has been difficult to establish a direct link between fracking and animal deaths (Bamberger, 2012). Another environmental problem arising from increased truck traffic is the introduction of new invasive species that endanger the habitat of existing species (Robbins, 2013).

According to a U.S. Government Accountability Study report (2012), there are 4,406 oil and gas wells in 105 National Wildlife Refuges of the nation. To stop environmental degradation from fracking, the U.S. Forest Service has decided to put extensive limits on fracking in Virginia's George Washington National Forest. This area holds the drinking water supply of Washington D.C. and serves as a major outdoor recreation area for Virginia's residents. Environmental groups in support of such action have called for the declaration of more frack free forest areas in the near future.

Faced with threats of regulations and under pressure from environmental groups, the oil and gas industry sometimes makes good faith promises to act as stewards of the environment, but their actual role in conservation is questionable. For example, the oil and gas industry had promised to protect the dune sagebrush lizard in western parts of Texas in its own interest, mainly to keep it off the federal endangered species list as drilling in any endangered species habitat is highly restricted. As a result, it created the Texas Habitat Conservation Foundation. Four years after the establishment of this foundation, a state review of its records revealed that it had failed to keep its promise of habitat restoration or monitor the activities of drillers and landowners to protect the lizards. As a result, the foundation was terminated by the Texas state comptroller in 2016 (Dexheimer & Price, 2016).

It is because of such occurrences that the public and environmental groups do not have much faith in private industries' promises to protect animal species or the environment. Therefore, the onus falls upon the states and the federal government to act as stewards of the environment. Unfortunately, the states' protective efforts are sporadic either due to absence of regulations or their weak enforcement. Whatever may be the case, they resent any federal intervention as it draws attention to their ineptitude in such matters. Nevertheless in 2016, the federal government's BLM declared 67 million acres of federal land as protected territory for the greater sage grouse, leaving behind only 3.5 million acres for cattle grazing, oil and gas drilling, and wind farms. The environmental groups have lauded this move, but it has annoyed many stakeholders with vested interests in such lands. Unsurprisingly, this prompted private groups representing the oil and gas industry, cattle ranchers, and alternative energy producers to file separate lawsuits against the Department of Interior (Osborne, 2016c).

Even after more than a decade of fracking there exists limited knowledge on the impacts of fracking fluid and wastewater spills on our fragile ecosystem. To fill the knowledge gap, scientists are willing to conduct independent studies, but lack of funding poses a serious obstacle. Other hindrances such as absence of baseline data to facilitate before and after comparisons, denial of access to private drilling sites for the collection of data, and the permitted non-disclosure of toxic chemicals used by oil and gas operators make it difficult to conduct scientific investigations. On the other hand, the findings of research conducted by the big oil and gas companies and funded with their own resources are not easily accessible to the public.

Community Concerns

The advent of fracking in communities has generated many safety and quality of life concerns, including traffic related fatalities, noise, and light pollution.

Vehicular Traffic

The movement of heavy trucks to and from the drilling site has become a major concern along with the hire of inexperienced drivers and use of vehicles that lack proper maintenance. On average, 2,300 to 4,000 trucks trips are required to deliver the drilling fluid to the well whereas the older practice of vertical drilling required one-third to one-half the number of truck trips. In the counties where fracking is being done, truckers and drivers of bulky equipment have been responsible for congestion of local streets and highways and deadly crashes that have killed many local residents and workers of the oil and gas industry. According to an Associated Press analysis of traffic related deaths and that of U.S. census data of six drilling states, fatalities have quadrupled here since 2004 despite an overall improvement in road safety and decline in numbers in other states. The truck drivers are equally at risk of losing their lives in accidents related to transportation of materials to and from the fracking sites. In one Texas drilling district, truck drivers faced the likely risk of dying in a fatal crash 2.5 times more per mile driven, in comparison with the state's average (Fahey & Begos, 2014).

A scrutiny of truck related problem reveals several loopholes in federal regulatory policy. The federal government's policy on the number of hours a truck driver can spend on the road is less stringent for the oil and gas industry's truck drivers. This has been made possible as a result of intense lobbying of Congress on behalf of the oil and gas industry for more flexibility in the time schedule of its truck drivers. For instance, if commercial truckers work 60 hours per week, they have to take off 34 hours to get two nights of sleep. The requirement for rest is much less for the oil and gas industry's truck drivers – they are required to take off only 24 hours for the same number of hours on the road. According to safety advocates, even though the exemptions do not apply to all types of trucks, like the pickup trucks and large tankers, oil and gas companies have been found to apply it to all trucks. It is also difficult for the Commercial Vehicle Safety Alliance, an association of police and highway authorities, to police the movement of trucks in the absence of a list from federal regulators of truck types that qualify for the exemption (Urbina, 2012).

In 2000, the federal highway authority tried to remove the exemption on grounds of safety concerns but failed in the face of opposition from the industry's lobbying groups. Attempts were made again in 2010 by federal

Table 5.5 Traffic Related Fatalities in Counties with Fracking Activities from 2009 to 2013 (%)

States	Increase or Decrease in Fatalities at Fracking Sites	Statewide Decline in Fatalities
North Dakota	+148	1
Texas	+18	20
Virginia	+42	8
New Mexico	−5	29
Pennsylvania	+4	19

Source: Associated Press, (2014). Deadly side effects to fracking boom.

authorities to amend highway regulations, but strong opposition from the trucking industry once again derailed such efforts (Urbina, 2012). To address the problem, some states have widened the roads and called for safer driving. At the industry level, safety programs have been adopted by companies, and truck traffic is being reduced through reliance on recycling programs and bringing water by pipelines to the drilling sites (Fahey & Begos, 2014). Despite these efforts, the ultimate decision to transport by trucks or pipelines depends on economics. Transport cost calculations have shown that it costs much more to transport by pipelines than trucks.

Additionally, there exists the problem of damage to infrastructure by heavy trucks and equipment. In 2012, the Texas Department of Transportation's Texas Energy Sector Roadway Needs taskforce identified the energy sector's trucking related damages to be worth $2 billion on the state's highways and an equal amount on county roads (Boske, Harrison, Moriarty, & McNew, 2014). In the Marcellus Shale of Pennsylvania, the consumptive use cost of state roadways of all types has been estimated at $13,000 to $23,000 per well or $5,000 to $10,000 per well if state roads with a lower traffic volume are excluded (Abramzon, 2014). Fracking's heavy toll on infrastructure in rural and urban areas calls for the review of policy options at the state level to make the oil and gas companies share more of the cost burden of repair and maintenance arising from the movement of truck traffic and heavy equipment to and from the drilling sites.

Noise and Light Impacts

At a fracking site, the noise from trucks and compressors has kept many people awake at night and poses a serious health risk to those individuals with high blood pressure, diabetes, and hearing loss. In addition, it has been blamed for loss in property value and individuals have filed lawsuits against oil and gas companies for the loss of enjoyment of their property (Goldberg, Williams, & Cours, 2015).

When fracking is conducted in isolated rural areas, landowners living at a distance from the drilling sites are not that much affected by the noise and other nuisances usually created by fracking. Conversely, in densely populated urban areas, fracking disturbs the lifestyles of many urban residents. In some areas, fracking related nuisances have given rise to tensions among neighborhood residents and energy producers. This is most evident in the Barnett Shale play, which stretches across 19 counties of North Texas and is populated by 1.6 million people. Some of these tensions have been diffused by producers who fear that residents' complaints might lead to obstructions in their drilling activities and cost them more money in the long run. Producers have either settled directly with local residents or through the cities or townships.

To cut down on noise pollution at production sites, the sound producing machines have been draped with sound blankets, which resemble large and heavy quilts. Cities and townships have also adopted remedial measures in the form of municipal codes to cut down on noise pollution at production sites. For example, cities like Fort Worth, Texas have put a limit on noise, which should not exceed more than five decibels during the day and three decibels at night. To comply with such codes, producers have moved their drilling sites closer to the roads where the noise levels are already much higher and there is also easy access to the sites (Kurth, Mazzone & Mendoza, 2012).

More complex cases of nuisance-based complaints have found their way to the courts for settlement by a jury. In these court cases, jury verdicts have been unpredictable, evident from three notable nuisance court cases in the Barnett Shale of North Texas. For example, in *Parr* v. *Aruba Petroleum*, the jury awarded the plaintiff $3 million upon their establishment that the company's operations affected not only the family's health but also led to a loss in property value and compelled the family to move away (Sadasivam, 2014). In the *Crowder* v. *Chesapeake Operating* case, even though the jury found the company's operations to pose a health risk from air pollution and disrupted the use of property through drilling noise, it only awarded $20,000 to the plaintiffs. In the *Teri Anglim* v. *Chesapeake Operating* case, similar to the Crowder case, the jury failed to find the company's activities to be a cause for nuisance. The variations in courts' responses to similar cases often leave the victims of fracking confused and disappointed (Hanen, 2015). Not knowing whom to complain to or how to recover their damages, many of them seek guidance from local non-profit agencies on how to navigate the bureaucracy of the state's regulatory agencies.

Additionally, birds and wild animals have also been affected by the continuous exposure to noise. The noise from fracking sites has brought about changes in their sleep patterns, foraging, anti-predatory behaviors, metabolism, and neuroendocrine, vascular, and reproductive systems (Barber et al., 2010; Kight, 2011). It is expected that such changes in animal behavior will affect the bio diversity and ecology of fragmented forests located at close proximity to fracking sites.

The bright electric lights used at drilling sites for safety purposes and the light from the flaring of natural gas also pose a serious threat to both humans and wildlife. At night, the bright lights help to light up the sky. From an aerial view of the night sky, the lights can be seen both in clusters and dispersed patterns. As seen in the picture below, a NASA image has captured the specs of bright light in the Eagle Ford Shale, which lies at close proximity to the nearby cities of Austin and San Antonio.

Miscellaneous Concerns

Other community concerns include various health related issues, increases in health care costs of those individuals living close to fracking sites and

Figure 5.5 The Lights Visible at Night from Fracking Sites in Texas.

Source: NASA, retrieved from http://earthobservatory.nasa.gov/IOTD/view.php?id=87725 &src=eoa-iotd.

where there also exists the risk of groundwater contamination (Muehlen-bachs, Spiller, & Timmins, 2015), and alteration of the natural aesthetics of the landscape with the placements of drilling rigs. In the tribal lands of American Indian communities, fracking has given rise to two groups as seen in the Bakken Shale of North Dakota. Here, one group of native tribe favors fracking for the hefty royalty payments they receive from the oil companies. The other group, with a deep spiritual relationship with land, is anti-fracking. They dislike the fumes emanating from the field sites and blame it for ozone depletion and asthma. Another group, the Omaha tribe of Nebraska, has protested against the extension of a pipeline from North Dakota into their territory. This native American community who once enjoyed the bounty from fracking in their territory also got to experience its ugly side when a pipeline carrying brackish water from a fracking site burst and spilled one million gallons into a ravine, killing plants and trees. As news of such incidents spread across the native Americans' territory, frack-ing met with more resistance from some groups. For example, an activist group of the Navajo tribe has joined hands with environmental groups in filing a federal lawsuit to block oil and gas drilling in their territory in New Mexico. The group is afraid that fracking might threaten their limited water supply and encroach upon their historic ruins (Osborne, 2016a).

A recent spate of studies has shown a correlation between health concerns and fracking related activities that release toxic chemicals. In absence of detailed investigations with substantial evidence, the oil and gas industry has ignored such claims. In the meanwhile, environmental groups have shown much interest in such studies and have expressed concerns over their findings. They are trying to create greater awareness and mobilize public support to bring about a change in affected communities. This is indeed not an easy feat when the political will to bring changes is lacking or weak in the state. As revealed by the history of the environmental movement, relevant changes in existing regulations and restrictions on the actions of the private industry sometimes require long drawn legal battles and perseverance. The same is true for fracking. Those individuals living close to fracking sites have to continue bearing the bulk of the burden of negative externalities both during and in the aftermath of fracking until the states are compelled to take adequate actions to address the problems.

Conclusion

The environmental impacts of fracking, especially on those individuals living close to fracking sites and on surrounding wildlife, warrant more attention. In many states, citizens' complaints of deterioration in the environment have prompted investigation and research by independent researchers at educational and environmental organizations. Their research findings help to lend credibility to public health concerns and justify the need for new and stringent regulations on fracking that will help to internalize the social costs of production. Even the EPA has been advised by its panel of scientists to reconsider its statement that fracking does not affect potable water quality in the face of new scientific evidences. It is true that the process of fracking per se may not contaminate water resources, but failures in inspection, maintenance, and accidents during transportation and at well sites do pose a significant risk. Irrespective of this risk, state governments in protecting oil and gas interests have decided either to disregard such findings or downplay their importance, citing insufficient evidence to link fracking to environmental damages. In the face of opposition from those people negatively impacted by fracking and to allay public fears, states have promised their own investigations. Since such investigations are costly and face bureaucratic hurdles in getting approved, people's choices are limited. They need either to be patient and silently suffer until funding for the study is approved or wait for an environmental organization to come to their rescue. In extreme cases, moving out of the area because of the environmental dangers is another extreme option that few would like to take.

Media focus and consequent increased awareness of environmental ills from fracking has made the oil and gas companies indulge in tactics to obscure the environmental woes or hide them behind the façade of economic gains. Repeatedly through various nationwide advertisement campaigns, the oil and gas industry has boasted of the many benefits of natural gas as a clean fuel and how it can fight climate change. The companies also point to how the overabundance in energy production is helping the nation make steady progress toward its goal of energy independence and have used jobs and profit figures as powerful bargaining tools to impede or delay regulatory actions. The lobbying groups working for the oil and gas industry continue to urge the government both at the state and federal levels not to make amendments on any existing regulations or introduce new ones. Such pressure tactics seem to work. The federal government has yet to remove the exemptions granted to the oil and gas industry in disposal of hazardous waste into underground injection wells that can contaminate groundwater, or in disclosure of chemicals used in fracking fluids.

At the state level, regulations on fracking are weak and non-existent in many aspects of this activity. Most states do not want to make the underground storage of wastewater expensive with the imposition of new fees or offer incentives to promote the greater use of recycled water in fracking, as observed in the state of Pennsylvania. Many states refuse to comply with certain federal regulations like the implementation of the green completion process, which has the potential to reduce methane emissions from field sites and therefore air pollution in respective areas. Further, states do not hesitate to file lawsuits against the federal government to reduce the cost of compliance of the oil and gas industry. Such actions partly reflect their lack of commitment to environmental protection and apathy toward the protection of people living close to fracking sites. Even in matters of wildlife protection, state regulations are either absent or if present, are weak and often lack enforcement. It is against such a backdrop of evidence that the jobs versus environment debate needs to be rekindled again. The various reports on fracking and its impacts on the economy and the environment can only evoke mixed reactions. Whatever stance one might take on the issue must require greater scrutiny of the issue along with separation of facts from fiction.

References

Abramzon, S., Samaras, C., Curtright, A., Litovitz, A., & Burger, N. (2014). Estimating the consumptive use costs of shale natural gas extraction on Pennsylvania roadways. *Journal of Infrastructure Systems, 20*(3), 014001.

Bamberger, M. & Oswald, R. (2012). Impacts of gas drilling on human and animal health. *New solutions: a journal of environmental and occupational health policy, 22*(1), 51–77.

Barber, J. R., Crooks, K. R., & Fristrup, K. M. (2010). The costs of chronic noise exposure for terrestrial organisms. *Trends in Ecology and Evolution, 25*(3), 180–189.

Barton, John. (2012). Presentation on TxDOT's energy sector task force. Retrieved from http://ftp.dot.state.tx.us/pub/txdot-info/energy/102312_txdot_presentation.pdf.

Boske, L., Harrison, R., Moriarty, B., & McNew, K. (2014). Potential use of highway rights-of-way for oil and natural gas pipelines. TxDOT 0-6581-Task 19-5. Transportation Policy Brief, *5*, 1–22. Retrieved from http://library.ctr. utexas.edu/ctr-publications/policy/0-6581_task19-5.pdf.

Brandt, A. R., Heath, G. A., Kort, E. A., O'Sullivan, F., Pétron, G., Jordaan, S., Tans, P., Wilcox, J., Gopstein, A., Arent, D., Wofsy, S., Brown, N., Bradley, R., Stucky, G., Eardly, D., & Harriss, R. (2014). Methane leaks from North American natural gas systems. *Science, 343*(6172), 733–735.

Bullard, R. D. (2000). *Dumping in Dixie: Race, class, and environmental quality* (Vol. 3). Boulder, CO: Westview Press.

Burton, G., Basu, N., Ellis, B., Kapo, K., Entrekin, S., & Nadelhoffer, K. (2014). Hydraulic "Fracking": Are surface water impacts an ecological concern? *Environmental Toxicology and Chemistry, 33*(8), 1679–1689.

Cooley, H. & Donnelly, K., (2012). Hydraulic fracturing and water resources: separating the frack from the fiction. Oakland, California: Pacific Institute.

Cook, M., Huber, K., & Webber, M. (2015). Who regulates it? Water policy and hydraulic fracturing in Texas. Texas Water Resources Institute, *Texas Water Journal, vol. 6*(1), 45–63.

Dexheimer, E. & Price, A. (2016). State fires organization formed by big oil to manage threatened lizard. Retrieved from www.mystatesman.com/news/news/state-fires-organization-formed-by-big-oil-to-mana/nqrgt/.

Elliott, D. (2016, August 16). Study: Most of methane hot spot in section of southwest U.S. is from natural gas leaks. *Houston Chronicle*, B7.

Energy Information Administration. (2016). Natural gas flaring in North Dakota has declined sharply since 2014. Retrieved from www.eia.gov/todayinenergy/detail.cfm?id=26632.

Eoh, Y. (2014). Yes, No, Maybe So: Uncertainty in Texas Groundwater Withdrawal for Hydraulic Fracturing. *Houston Law Review, 52*, 1227.

EPA. (2011). *Plan to Study the Potential Impacts of Hydraulic Fracturing on Drinking Water Resources*. Washington, D.C.: Office of Research and Development. Retrieved from www.epa.gov/hfstudy/HF_Study__Plan_110211_FINAL_508.pdf.

Fahey, K. & Begos, K. (2014). AP Impact: Deadly side effect to fracking boom. Retrieved from http://bigstory.ap.org/article/ap-impact-deadly-side-effect-fracking-boom-0.

Fisher, M., Mayer, A., Vollet, K., Hill, E., & Haynes, E. (2018). Psychosocial implications of unconventional natural gas development: Quality of life in Ohio's Guernsey and Noble Counties. *Journal of Environmental Psychology, 55*, 90–98.

Fontenot, B., Hunt, L., Hildenbrand, Z., Carlton Jr, D., Oka, H., Walton, J., Hopkins, D., Osorio, A., Bjorndal, B., Qinhong, H., & Schug, K. (2013). An evaluation of water quality in private drinking water wells near natural gas extraction sites in the Barnett Shale Formation. *Environmental Science & Technology, 47*(17), 10032–10040.

Fox, J. (2010). Gasland, a 2010 documentary film on fracking, nominated for Academy award for best documentary film in 2011.

Fragoso, A. (2016). The EPA hasn't updated fracking rules in nearly 3 decades. Now, environmental groups are suing. Think Progress. Retrieved from http://thinkprogress.org/climate/2016/05/05/3775544/environmentalists-sue-epa-for-drilling-waste/.

Freyman, M. (2014). Hydraulic fracturing and water stress: water demand by the numbers. A Ceres Report.

Frohlich, C., Hayward, C., Stump, B., & Potter, E. (2011). The Dallas–Fort Worth earthquake sequence: October 2008 through May 2009. *Bulletin of the Seismological Society of America, 101*(1), 327–340.

Fry, M., Briggle, A., & Kincaid, J. (2015). Fracking and environmental (in)justice in a Texas city. *Ecological Economics, 117*, 97–107.

Galbraith, K. (2013, March 13). Ambiguities Reign in Regulations for Groundwater Fracking. *Texas Tribune*. Retrieved from www.texastribune.org/2013/03/13/fracking-groundwater-rules-reflect-legal-ambiguiti/.

Geiling, N. (2016). EPA's Efforts to curb methane emissions suffers a setback as 13 states sue. *Think Progress Climate News*. Retrieved from http://thinkprogress.org/climate/2016/08/03/3804612/methane-rule-epa-states-lawsuit/.

Goldenberg, S. (2013). A Texan tragedy: ample oil, no water. Retrieved from www.theguardian.com/environment/2013/aug/11/texas-tragedy-ample-oil-no-water.

Goldberg, H., Williams, M., & Cours, D. (2015). It's a Nuisance: The Future of Fracking Litigation in the *Wake of Parr* v. *Aruba Petroleum, Inc., Virginia Environmental Law Journal, 33*(1), 1–22.

Groat, C. & Grimshaw, T. (2012). Fact-Based Regulation for Environmental Protection in Shale Gas Development. The Energy Institute, University of Texas at Austin. Austin, Texas.

Hanen, K. (2015). Nuisance easements: the new cost of doing business in the oil and gas industry. Retrieved from http://tjogel.org/nuisance-easements-the-new-cost-of-doing-business-in-the-oil-and-gas-industry/.

Hildenbrand, Z. L., Carlton Jr, D. D., Fontenot, B. E., Meik, J. M., Walton, J. L., Taylor, J. T., Thacker, J., Korlie, S., Shelor, C., Henderson, D., Kadjo, A., Roelke, C., Hudak, P., Taylour, B., Rifai, H., & Schug, K. (2015). A comprehensive analysis of groundwater quality in the Barnett Shale region. *Environmental science & technology, 49*(13), 8254–8262.

Hiller, J. (2016, May 16). Quakes, drilling long intertwined in Texas. *San Antonio Express News*, B1.

Hirji, Z. (2016). Oil and gas quakes have long been shaking Texas, new research finds. *Inside Climate News*. Retrieved from https://insideclimatenews.org/news/17052016/oil-and-gas-production-earthquakes-texas-new-research-fracking.

Hornbach, M., DeShon, H., Ellsworth, W., Stump, B., Hayward, C., Frohlich, C., Oldham, H., Olson, J., Magnani, B., Brokaw, C., & Luetgert, J. (2015). Causal factors for seismicity near Azle, Texas. *Nature Communications*, 6.

Jackson, R., Pearson, B., Osborn, S., Warner, N., & Vengosh, A. (2011). Research and Policy Recommendations for Hydraulic Fracturing and Shale-Gas Extraction. Center on Global Change, Duke University, Durham, NC.

Johnston, J. E., Werder, E., & Sebastian, D. (2016). Wastewater Disposal Wells, Fracking, and Environmental Injustice in Southern Texas. *American Journal of Public Health, 106*(3), 550–556.

Jones, E. (2016). EPA releases first ever standards to cut methane emissions from the oil and gas sector. EPA News Release. Retrieved from www.epa.gov/news releases/epa-releases-first-ever-standards-cut-methane-emissions-oil-and-gas-sector.

Kahrilas, G., Blotevogel, J., Stewart, P., & Borch, T. (2014). Biocides in hydraulic fracturing fluids: a critical review of their usage, mobility, degradation, and toxicity. *Environmental Science & Technology, 49*(1), 16–32.

Kight, C. R. & Swaddle, J. P. (2011). How and why environmental noise impacts animals: an integrative, mechanistic review. *Ecology letters, 14*(10), 1052–1061.

Konikow, L. F. (2015). Long-Term Groundwater Depletion in the United States. *Groundwater, 53*(1), 2–9.

Kulander, C. S. (2013). Shale oil and gas state regulatory issues and trends. *Case Western Reserve Law Review, 63*, 1101–1140.

Kurth, T. E., Mazzone, M. J., & Mendoza, M. S. (2012). American Law and Jurisprudence on Fracing—2012. *Rocky Mountain Mineral Law Foundation Journal, 58*(4).

Lavelle, M. (2017). Trump's EPA halts request for methane information from oil and gas producers. *Inside Climate News.* Retrieved from https://inside climatenews.org/news/03032017/scott-pruitt-environmental-protection-agency-methane-greenhouse-gas-climate-change.

Malewitz, J. (2016, August 24). Texas promised 34 years ago to track oilfield waste in aquifers. It didn't. *Texas Tribune.* Retrieved from www.texastribune.org/2016/08/24/texas-promised-34-years-ago-track-oilfield-waste-a/.

Malewitz, J. (2015, November 3). On quakes, regulator sides with oil companies. *Texas Tribune.* Retrieved from www.texastribune.org/2015/11/03/quake-question-railroad-commission-sides-energy-co/.

May, J. (2012). The increasing burden of oil refineries and fossil fuels in Wilmington, California and how to clean them up! A report from Communities for Better Environment. Retrieved from www.cbecal.org/wp-content/uploads/2012/05/wilmington_refineries_report.pdf.

McDermott-Levy, R., Kaktins, N., & Sattler, B. (2013). Fracking, the environment, and health. *The American Journal of Nursing, 113*(6), 45–51.

Mckenzie, L. M., Witter, R. Z., Newman, L. S., & Adgate, J. L. (2012). Human health risk assessment of air emissions from development of unconventional natural gas resources. *Science Total Environment, 424*, 79–87.

Miller, K. (2016, September 4). Record-tying Oklahoma earthquake shatters nerves over widespread area. *Houston Chronicle*, A3 & A4.

Miller, S. M., Wofsy, S. C., Michalak, A. M., Kort, E. A., Andrews, A. E., Biraud, S., Dlugokencky, E., Eluszkiewicz, J., Fischer, M., Janssens-Maenhout, G., Miller, B., Miller, J, Montzka, S., Nehrkorn, T., Sweeney, C., & Miller, B. (2013). Anthropogenic emissions of methane in the United States. *Proceedings of the National Academy of Sciences, 110* (50), 20018–20022.

Meng, Q. & Ashby, S. (2014). Distance: A critical aspect for environmental impact assessment of hydraulic fracking. *The Extractive Industries and Society, 1*(2), 124–126.

Moore, C., Zielinska, B., Pétron, G., & Jackson, R. B. (2014). Air impacts of increased natural gas acquisition, processing, and use: a critical review. *Environmental science & technology, 48*(15), 8349–8359.

Mitchell, A. L., Tkacik, D. S., Roscioli, J. R., Herndon, S. C., Yacovitch, T. I., Martinez, D., Vaughn, T., Williams, L., Sullivan, M., Floerchinger, C., Omara, M., Subramaniam, R., Zimmerle, D., Marchese, A., & Robin, A. (2015). Measurements of methane emissions from natural gas gathering facilities and processing plants: Measurement results. *Environmental science & technology, 49*(5), 3219–3227.

Muehlenbachs, L., Spiller, E., & Timmins, C. (2015). The housing market impacts of shale gas development. *The American Economic Review, 105*(12), 3633–3659.

Nall, R. (2013). Silicosis. A blog posting from Healthline. Retrieved from www. healthline.com/health/silicosis#Overview1.

National Institute of Occupational Safety and Health Administration. (2012). Hazard alert: worker exposure to silica during hydraulic fracturing. Retrieved from www.osha.gov/dts/hazardalerts/hydraulic_frac_hazard_alert.pdf.

Osborne, J. (2016a, September 18). Tribes split on drilling: Despite protests, some are drawn to big oil profits. *Houston Chronicle*, A1 & A25.

Osborne, J. (2016b, August 2). Texas sues EPA over tougher rule on methane leaks. *Houston Chronicle*, B1.

Osborne, J. (2016c, July 1). Oil firms' battle with sage grouse rises again. *Houston Chronicle*, A1.

Osborne, J. (2016d, March 29). Quake risk rises in Texas, Oklahoma. *Houston Chronicle*, A1 and A8.

Occupational Safety and Health Administration. (2016). OSHA's final rule to protect workers from exposure to respirable crystalline silica. Retrieved from www.osha.gov/silica/.

Percival, P. (2010a). TCEQ concerned about emissions asks industry to help find Barnett Shale facilities and fix emissions problems. *Basin Oil and Gas* magazine blog post. Retrieved from www.slideshare.net/ppercival/magazine-clips-percival.

Percival, P. (2010b). Service companies offer help with emission surveys, mitigation efforts. *Basin Oil and Gas* magazine blog post. Retrieved from www.slideshare. net/ppercival/magazine-clips-percival.

Railroad Commission of Texas. (2016). Flaring Regulation. Retrieved from www. rrc.state.tx.us/about-us/resource-center/faqs/oil-gas-faqs/faq-flaring-regulation/.

Rawlins, R. (2013). Planning for fracking on the Barnett shale: Urban air pollution, improving health based regulation, and the role of local governments. *Virginia Environmental Law Journal, 31*(226), 226–306.

Robbins, K. (2013). Awakening the slumbering giant: How horizontal drilling technology brought the endangered species act to bear on hydraulic fracturing. *Case Western Reserve Law Review, 63*(4), 13–16.

Robinson, A. (2012). Air pollutant emissions from shale gas development and production. Slides. *Pittsburgh (PA): Carnegie Mellon University.*

Roy, A., Adams, P., & Robinson, A. (2014). Air pollutant emissions from the development, production, and processing of Marcellus Shale natural gas. *Journal of the Air & Waste Management Association, 64*(1), 19–37.

Rundquist, S. (2014). Danger in the air. A blog posting from the Environmental Working Group (EWG). Retrieved from www.ewg.org/research/danger-in-the-air.

Sadasivam, N. (2014). Aggressive tactic on the fracking front. Pro Publica. Retrieved from www.propublica.org/article/aggressive-tactic-on-the-fracking-front. Accessed on 6/17/2016.

Schmidt, C. (2013). Estimating wastewater impacts from fracking. *Environmental Health Perspectives, 121*(4), A117.

Schmidt, C. (2011). Blind Rush? Shale gas boom proceeds amid human health questions. *Environmental Health Perspectives, 119*(8), A348–53.

Song, L., Morris, J., & Hasemyer, D. (2014). Fracking boom spews toxic air emissions on Texas residents. Inside Climate News. Retrieved from https://insideclimatenews.

org/news/20140218/fracking-boom-spews-toxic-air-emissions-texas-residents. Accessed on 8/3/2016.

Theodori, G. (2012). Public perception of the natural gas industry: Data from the Barnett Shale. *Energy Sources, Part B: Economics, Planning, and Policy, 7*(3), 275–281.

Tomlinson, C. (2016, September 9). Texas can learn from Sooner State on quakes. *Houston Chronicle*, A1 & 15.

Tubb, L. (2010). Childhood asthma: a guide to action. The results of an asthma think tank conducted on December 10, 2010 by The Center for Children's Health, Cook's Children Healthcare System.

U.S. Department of the Interior. (2001). Technical Measures for the investigation and mitigation of fugitive methane hazards in areas of coal mining. Office of Surface Mining. September 2001. Print.

U.S. Government Accountability Office. (2016). Oil and gas: interior could do more to account for and manage natural gas emissions. Retrieved from www.gao.gov/assets/680/678285.pdf.

U.S. Government Accountability Office. (2012). Oil and gas: Information on shale resources, development, and environmental and public health risks. Report to congressional requesters. Washington, D.C.

U.S. Government Accountability Office. (2010). Federal oil and gas leases: opportunities exist to capture vented and flared natural gas, which would increase royalty payments and reduce greenhouse gases, GAO-11-34, Washington, D.C. Retrieved from www.gao.gov/new.items/d1134.pdf.

U.S. Government Accountability Office. (2004). *Natural Gas Flaring and Venting: Opportunities to Improve Data and Reduce Emissions*, GAO-04-809, Washington, D.C.

Urbina, I. (2012). Deadliest danger isn't at the rig but on the road. *New York Times* online, May 14, 2012. Retrieved from www.nytimes.com/2012/05/15/us/for-oil-workers-deadliest-danger-is-driving.html?pagewanted=3&_r=0.

Vidic, R., Brantley, S., Vandenbossche, J.,Yoxtheimer, D. & Abad, J. (2013). Impact of shale gas development on regional water quality. *Science, 340*, 1235009-01 to 1235009-09.

Weinhold, B. (2012). The future of fracking: new rules target air emissions for cleaner natural gas production. *Environmental Health Perspectives, 120*(7), A272.

Wethe, D. (2016, August 15). Sand sellers poised to pile up profits: oil producers find the more grit the better when fracturing. *Houston Chronicle*, Business section, B1 and B3.

Wittmeyer, Hannah. (2013). Fracking regulations in Texas. Retrieved from http://frackwire.com/texas-regulations/.

Zelinski, A. (2016, September 10). Paxton is rejoining climate change fray: 11 AGs are suing to block probe by Massachusetts. *Houston Chronicle*, A1 and A17.

6 Case Studies on Communities Oppositions to Fracking

Introduction

At the time of adoption of the fracking technology, its externalities were unknown. Over the years with the proliferation of fracking in the state, the externalities have become evident and can no longer be ignored. The oil and gas companies try hard not to acknowledge them and overlook these externalities in their portrayal of this advanced drilling procedure as the harbinger of energy abundance and economic prosperity in the twenty-first century. The lack of attention given to these externalities by the industry and the state and local governments has led to inadequate protection of those individuals who live in communities closer to the drilling sites. They are the most susceptible to the adverse impacts of the externalities during and in the aftermath of fracking. Their frustrations have evoked community opposition to fracking, which have varied from intense to mild ones as observed in the four case studies presented in this chapter.

In selection of each case study to investigate a community's response to fracking, media focus has played an important role. Those cases that received adequate attention in local, state, and national newspapers were chosen for further analysis. In writing the case studies, information has been retrieved from multiple sources, including newspapers, books, blogs, community websites, and non-profit organizations. Additionally, interviews with community residents, local public officials, and members of environmental organizations have helped enrich the details in each case and make them more robust.

Each case study included in the chapter focuses on an externality that has impacted individuals' quality of lives or their sense of wellbeing in a community and how they have responded to it. These externalities range from air pollution and induced earthquakes to water shortages and various transportation related issues. For example, the city of Denton experienced air pollution problems when fracking was at its peak in the community while the city of Irving faced a series of human induced seismicity that disturbed the peace of mind of its residents. In Carrizo Springs, though fracking took place outside the city's boundary, it consumed a lot of water,

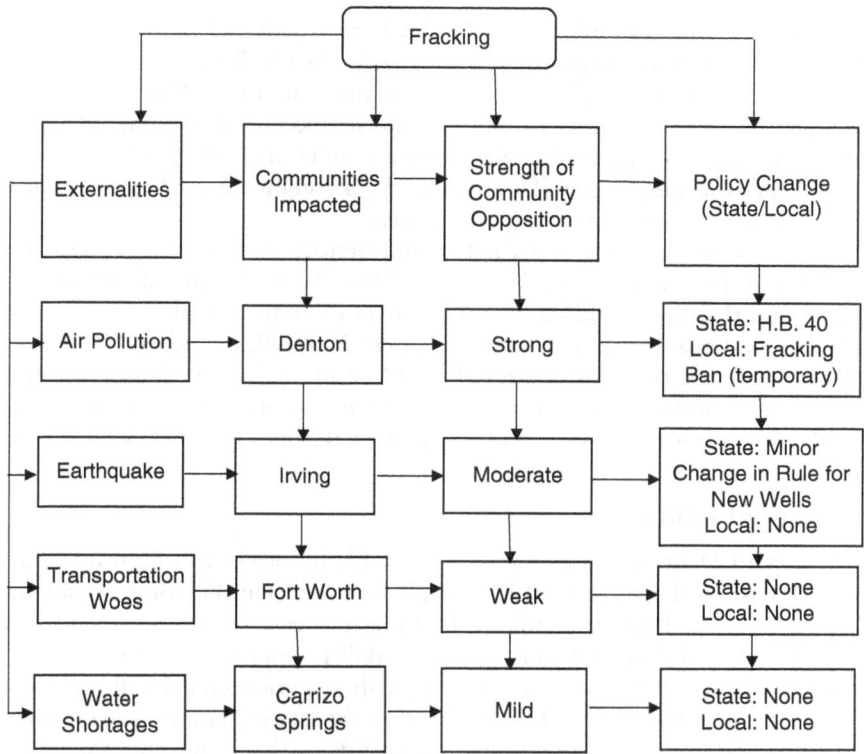

Figure 6.1 A Conceptual Framework of Communities Opposition to Fracking and
Policy Outcomes.

leading to shortages in agriculture whereas in the city of Fort Worth, the
movement of heavy trucks back and forth from the drilling sites left indeli-
ble damages to its infrastructure of roads and highways. Collectively, the
damages from externalities warrant attention not only from decision
makers and politicians but also planners and the public, who have the
potential and authority to take various issues under consideration and
enable relevant policy changes in the existing regulatory framework of the
state and local governments.

Typical of any case study, the cases offer a lesson on the success or
failure of a community's opposition to fracking through descriptions and
narration of various events, processes, and responses of individuals. Simultaneously, they also offer the readers the scope to judge for themselves the
validity of a community's opposition to fracking and determine the need, if
any, for policy changes that would only cost the oil and gas companies a
lot of money.

The Case of Denton and its Fracking Ban

Lying at the northern edge of the Dallas Fort Worth metro area, the city of Denton is one of the fastest growing cities in North Texas. It is a medium-sized "home ruled city," that is it can manage its own affairs in the best interest of its residents with minimal interference from the state. The city has a land area of approximately 89 square miles and is located in a county that bears the same name. Due to the city's somewhat central location in the county, it also serves as the county seat.

The city is the home of two state universities and they have helped to build its reputation as a college town. These two educational institutions dominate the city's landscape and serve as its major employers. In 2016, the city had a population of approximately 123,000, including 50,000 students from the two universities (City of Denton, 2016). With its educational institutions, two hospitals, a vibrant downtown, and parks and trails, this is one of the fastest growing cities in the state and the nation.

History of Denton

The city of Denton was established in 1857. Its origin can be traced back to a donation of 100 acres of land made by three Denton County residents, who played an important role in building the city. In 1866, the city was incorporated and at that time it had a small population of 361 people. In the first decade of its existence, the city only experienced a small growth in population. It was not until the 1880s that the city experienced real growth in its population with the expansion of railways that connected the city to the north and the south. Since then the city's population growth has continued into the twenty-first century.

In addition to its status as a county seat, the city also served as an agricultural trade center in the past. It had mills that processed agricultural products like wheat, corn, and cottonseed. In 1890, the city was chosen as the site for the establishment of the North Texas Normal College, later renamed the University of North Texas (UNT). After a decade, another educational institution was established in 1903 – the Girls' Industrial College (Odom, 2016), which is currently known as Texas Woman's University.

Drilling in Denton

The city of Denton lies on top of the Barnet Shale and it was only 37 miles south of the city that fracking first started in the city of Fort Worth, at the beginning of the twenty-first century. As fracking gained momentum in the Barnett Shale, it gradually spread to other cities located on the same shale formation, known for its natural gas reserves. Fracking made its debut in Denton only a few years later after it started in Fort Worth and was added to the city's list of numerous business activities.

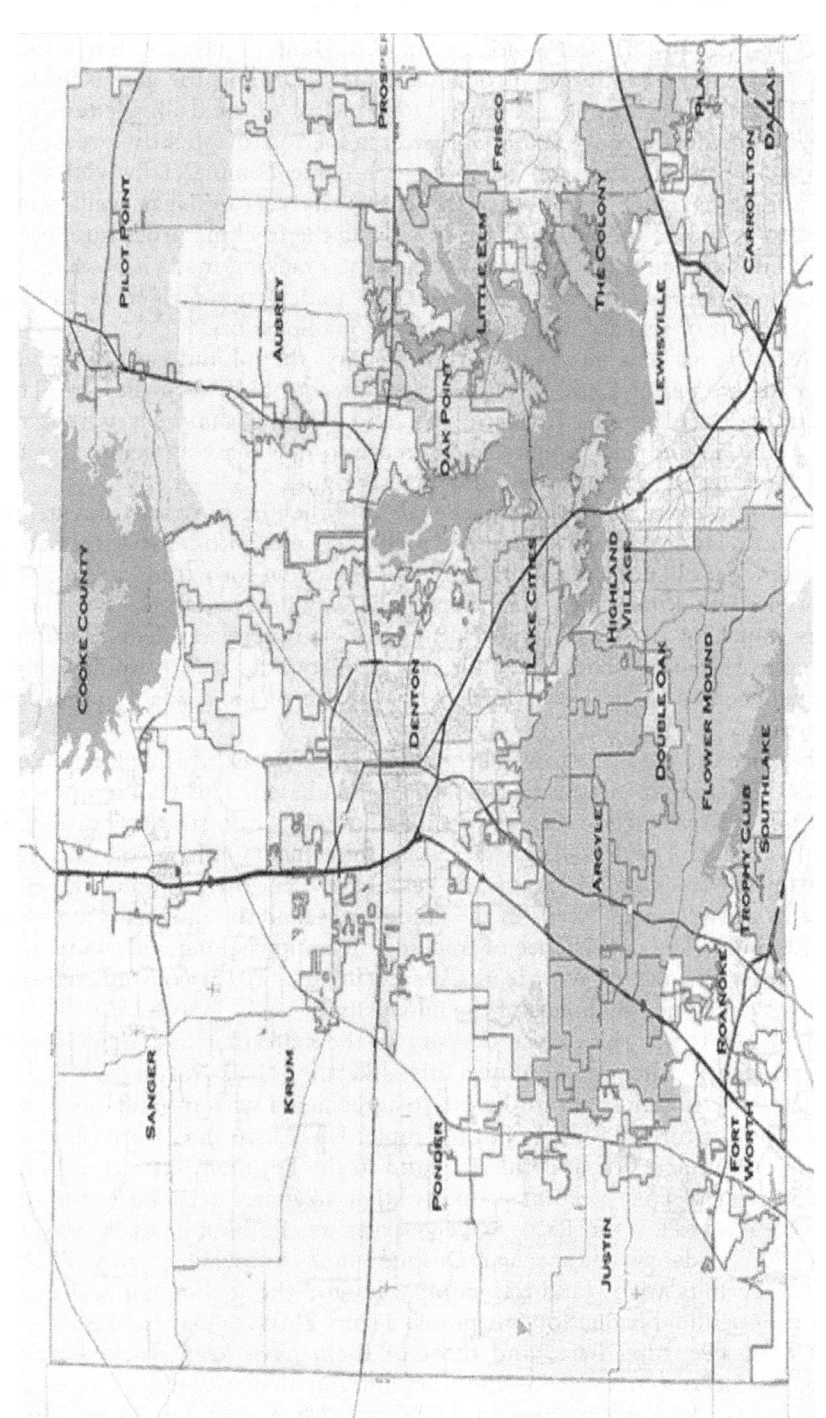

Figure 6.2 Location of the City of Denton in North Texas.

Source: Denton County, Texas.

Oil and gas production is nothing new in Denton. The city has a long history of energy production from fossil fuels. Drilling for gas using the traditional method started as early as 1949. Most of the drilling sites were located outside the city limits but after a period of slightly over four decades, when the fracking technology became commercially viable in 1997, fracking made its appearance in the city. Natural gas wells were drilled at both old and new drilling sites using the fracking procedure, both within and outside the boundaries of the city. Fracking made it possible to extract natural gas from the deep seated underground reserves in the northern part of the city, which was unimaginable before.

In drilling for new gas wells within the city, the oil and gas companies had to secure permits not only from the state but also from the city. The city granted the permits to frack new and existing gas wells within its limits. The oil and gas companies entered into new lease agreements with those landowners who also held the mineral rights.

The discovery of natural gas in the city and the lure of royalty payments drew many landowners with mineral rights to enter into lease agreements with the local oil and gas companies. Soon, fracking for natural gas picked up momentum within the city and made its way into residential neighborhoods, much to the surprise and dismay of many city residents. Drilling rigs appeared near schools, children's playgrounds, and hospitals, and fracking seemed to pose a hazard to the health of individuals in various age groups.

The once quiet college town succumbed to the pressure of drilling activities and fracking for natural gas continued unabated. This proliferation of fracking sites within a college town may sound a little strange but it did actually occur. The UNT signed a lease agreement with an oil and gas company for fracking on campus. It held 33 percent ownership of mineral rights from two pooled wells in the land just behind the university's newly built stadium. The coexistence of fracking on campus along with wind turbines and a slogan of "We Mean Green" (Briggle, 2015) contradicted the environmental friendly image of the university.

The UNT is not the only university in the state that had signed such an agreement. Other public universities like the UT at Austin and Texas A & M had also entered into lucrative agreements with oil and gas companies long before the UNT. For example, UT Austin has leased half of its 2.1 million acres of mineral rich land in the Permian Basin to oil and gas companies. This piece of property, that stretches across 19 counties and is traversed by the Pecos River, serves as the habitat of migratory birds and endangered species. Despite such knowledge, the TRRC granted permits to oil and gas companies for the drilling of wells for exploration and production purposes. From 2005 onwards, 4,350 gas wells have been dug here, and most of them have been fracked using scarce water resources from the surrounding drought prone areas of West Texas. Accidental spills in 2008 and 2009 polluted the soil and groundwater in some parts of the leased property (Inglis & Metzger, 2015). Such incidents have been dismissed by university officials as equivalent to the emptying of "one can of Coke per acre per year" over a period of eight years (Fox, 2015).

Gas Wells in Denton

In a state climate that is conducive to fracking, gas wells have been drilled just behind the backyard fence of houses in some residential neighborhoods of Denton. Although there exists no record of the exact year and date when fracking started in the city, the city's first ordinance on gas wells dates back to 2001, so the assumption is that it may have started during that period. Currently, there exist 277 active gas wells within the corporate city limits (COD) and another 225 active gas wells (City of Denton, 2016) in its extraterritorial jurisdiction (ETJ). See Figure 6.3A.

The city has also entered into lease agreements with oil and gas companies. It owns mineral rights on public lands and has availed itself of the opportunity to earn revenue from its leases. For the fiscal year of 2015 to 2016, it earned a revenue of $352,501 for its Airport Gas Fund and $82,488 for its Parks Gas fund. Much of that money has been spent on airport and park improvement projects within the city to benefit its residents.

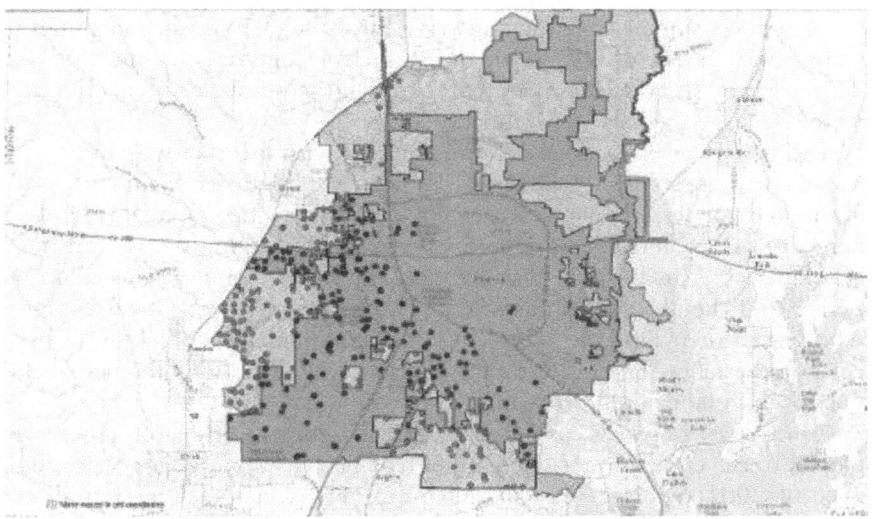

Figure 6.3A Active Gas Wells In and Around the City of Denton.

Source: City of Denton.

Note
Red dots indicate gas wells within COD and green dots represent gas wells in the ETJ.

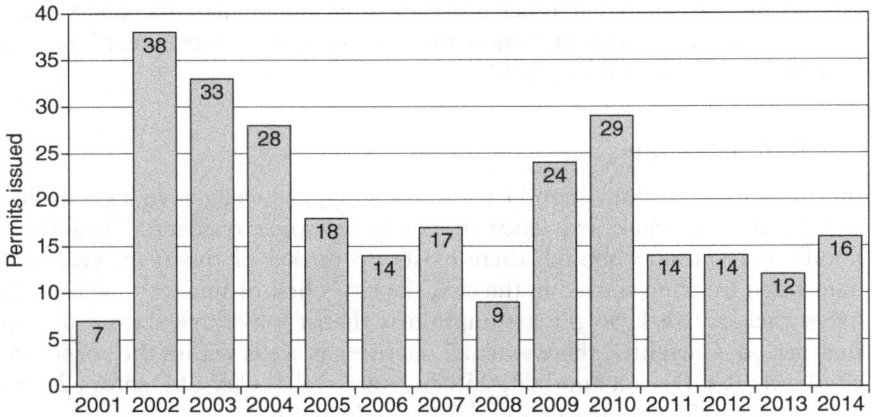

Figure 6.3B Number of Gas Well Permits Issued by Year in the City Limits of Denton.

Source: City of Denton

Fracking in the City

The adoption and rapid spread of fracking in the state was fueled by the desire for an increase in energy production and generation of revenue from it. The state's support for fracking had helped the oil and gas companies to secure permits for drilling from the Railroad Commission, but city residents of Denton failed to support fracking in their backyard when permits were issued by the local government.

The actual revenue earned from fracking within the city amounted to less than 1 percent of the city's budget, while taxes from gas wells accounted for less than half a percent of property taxes. Further, only 2 percent of Denton residents owned mineral rights within the city. The main beneficiaries of fracking in the city were the out of state oil and gas companies and the absentee landowners. The landowners with mineral rights who leased their lands to the oil and gas companies for fracking did not get to experience the negative externalities that were the outcomes of this unconventional drilling procedure.

From 2010 onwards, as fracking gained momentum in the city, complaints to the city from residents mounted. The city's initial gas well ordinance of 2001 required a small setback of 250 feet and it permitted the location of gas wells both in residential and mixed zoned areas. At the time of the passage of the initial ordinance, the city residents did not bother about the setback requirements since most of the gas wells were mostly located outside the city limits in the ETJ. It was when fracking started making inroads into residential neighborhoods that some residents found themselves living too close to gas wells (Friedman & Price, 2016). Other people also found themselves in a similar predicament when gas wells were sited close to a senior living center, hospital, children's park, and elementary and high schools.

The drilling and production process using the fracking technology created many nuisances that were unforeseen or unheard of in a quiet university town like Denton. At the drilling and production sites, the idling of trucks on an average of six hours per trip contributed much to the air pollution. This posed a problem because neither the city nor Denton County had any restrictions on idling of trucks unlike the neighboring counties of Tarrant, Collin, Dallas, and Kaufman.

The deterioration in air quality in the city and the county made the American Lung Association (ALA) give the county a failing or "F" grade in air quality in its State of the Air quality report. Since 2011, Denton County had failed to meet the eight hours standard for ozone and had received a failing or "F" grade in air quality. According to the ALA, the county's number of high ozone days had exceeded that of nearby Dallas County, which was also located in an ozone non-attainment area. As a result, Denton County was listed with those 20 to 25 counties in the state that had a relatively high number of ozone alert days every year.

Other problems from fracking included noise and light pollution that caused disturbances in sleep patterns among those residents whose houses were located close to the fracking sites. In two residential neighborhoods close to the fracking sites, 25 homeowners filed a lawsuit against Eagleridge Energy and the Eagleridge Operating Company in 2014. The residents claimed that fracking not only led to a decrease in their property value but also limited the use of their property. For example, the noise, dust, and offensive odor in the air prevented residents from spending time outdoors (Hunt, 2014). Even the increase in truck traffic within the city added to the traffic congestion and safety concerns on local roads and highways.

Table 6.1 Denton County's High Ozone Days and Air Quality Ranking

Year	High Ozone Days (weighted average)	Rank	Grade of Air Quality
2016	26.7	22	F
2015	21.7	14	F
2014	17.3	22	F
2013	17.3	16	F
2012	10.7	Unavailable	F
2011	12.5	Unavailable	F

Source: American Lung Association, State of the Air Report.

Note
F grade refers to 9 or more unhealthy ozone days over the standard.

The disturbances from fracking in the residential neighborhoods and the deterioration in the quality of life of residents posed serious concerns among many. They not only filed complaints with the city's office but also went as far as contacting their local representatives and the Railroad Commission's office and requested an end to fracking near their residences, schools, and parks. All such efforts proved futile. The respective offices reminded the city residents to contact their city office instead since they had little or no jurisdiction over their issues in the city.

According to a city resident, when complaints about noxious odors that polluted the air were filed with the city, the city did send a police officer to check the production site. The officer would later contact the complainant to inform them that nothing could be done about it since the operator had the approval or the permit from the city. Frustrated with such answers, some concerned residents even went as far as lodging their complaints with the office of the TCEQ. The office did send an inspector to the fracking site a few days later but the noxious odor had disappeared by then. In such a scenario, neither the inspector nor the police officer could do anything to alleviate the residents' problems.

In another event in 2009, when homeowners saw a Fort Worth-based gas company, Range Resources, inspecting a vacant piece of land to drill gas wells in Rayzor Ranch, they became concerned and contacted city officials. Fearing that gas wells would be sited too close to their homes, a hospital, retirement center, and a park, the residents requested the relocation of the drilling site to another area of the same property. People in the community led a petition drive and collected almost 200 signatures from home owners to request the city to stop the approval of the permit. One doctor, who had earlier been denied a permit by the city to build a doctor's office in the area, collected 100 signatures from his colleagues to petition the city to deny the energy company a permit. Unfortunately, the city failed to respond to the citizens' request and the permit was granted.

In issuance of the permit, the city council members listened to the residents' testimony and reviewed the scientific evidence on the health impacts of fracking that were submitted by the residents. After taking all those things into consideration, along with the fact that the city itself had sued the same company for $400,000 for unpaid royalties, the city council nevertheless voted six to one in favor of issuance of the permit. The concerned residents attending the meeting were later told that they would have the city council's support if they wanted to take the issue to the state government.

City residents present at the city council meeting knew that the city council's decision to issue the permit was driven by their fear of costly litigation with the gas company. One of the council members who voted in favor of the permit later admitted that in casting his vote, he felt a gun was pointed to his head. The representative from the oil and gas company that leased the parcel of land argued at the hearing that the particular spot (close to the residential homes) was chosen because it had the most potential of yielding the maximum amount of natural gas at that site. Moving the drilling site to another location would not yield the same result and was therefore not feasible. He also reminded the people present at the meeting that any refusal to grant a permit would amount to a deprivation of the owner's mineral rights (Malewitz, 2014b).

As fracking continued unabated in the city, other risks appeared in the community. In April 2013, there was a blow out of a gas well near the city's airport that put Denton residents, businesses, and the local airport at risk. The blow out occurred at 1:30 am but was not reported to the Denton Fire Department until 10:45 am by the Eagle Ridge company, the operator of the well. Upon investigation, it became clear there was a delayed response to the incident, which posed a risk of flash fire in Denton. When the news reached the city and state officials, local policemen, firefighters, and well control specialists rushed to the well site to contain the fire.

Toxic chemicals were released into the air at the well site during the fire. To identify the chemicals, samples were taken a little before and after the well fire was brought under control at 3:39 pm. A 30 minutes downwind air quality sample collected before well closure showed evidence of 46 of 84 hazardous chemicals, including benzene and ethylene dibromide, that are carcinogenic. An upwind sample collected after the well closure helped to detect 27 of the 84 hazardous chemicals (Heinkel-Wolfe, 2013).

The City's Responses to Residents' Complaints

The city officials, in response to residents' numerous complaints on fracking, ordered the construction of walls surrounding the gas wells, through the gas well ordinance. However, these walls proved to be inadequate in addressing the problem. Though the barrier walls succeeded in hiding the production sites from the eyes of the people, they failed to contain the sounds or noxious fumes that emanated from the enclosed sites. The air pollution and loud noise posed health concerns to residents and downgraded their quality of life. It discouraged adults and children alike from stepping outside their houses and some parents noticed an increase in nose bleeds and respiratory problems in their children.

The only tool that was available to the city to control fracking was the local ordinance. The initial gas drilling and production ordinance now became the subject of concern. It was revised several times due to environmental and land incompatibility concerns. The first wave of changes was made during the 2002 to 2004 period followed by another wave in 2010. Despite such changes, many residents still found themselves in very close proximity to the gas wells. Although the revised ordinance required a setback of 500 feet for gas wells in residential neighborhoods, the old 250 feet setback requirement still prevailed if the permits were secured before the revision was made (Fry, Briggle, & Kincaid, 2015).

Figure 6.4 A Wall Built Around the Gas Drilling Site in a Residential Neighborhood of Denton.

Source: Author.

The grandfathering of the ordinance's clause did not help those city residents who lived in residential neighborhoods with preexisting well pad
sites. These individuals expressed concerns over their personal welfare and
safety issues. Their intensification of protests against fracking made the
city issue a moratorium on gas well permits for fracking in February 2012
for 120 days. It was during the moratorium period that the city reviewed
its gas well ordinance again.

In 2012, before the expiration of the moratorium, a task force comprising five voting members with two representatives from the shale gas
industry and three local residents (one of whom was a retired petroleum
engineer) was given the responsibility to make recommendations for
amendments in the ordinance. At the end of the review process, a three to
two majority voted in favor of the energy industry and called for the elimination of the city's strict rules on noise, well casings, public notifications,
and other industrial activities (Brown, 2012).

In 2013, a revised version of the ordinance with a setback requirement
of 1,200 feet went into effect for new gas wells. Nine months after enforcement of the new requirements, homeowners in another residential neighborhood protested against fracking. Here, a gas well operator sited three
wells close to residential homes. Since the existence of pad sites predated

that of the particular residential development in the city, the operator claimed he had the vested rights in siting the new wells in an old pad site. These new wells were sited only 250 feet away from their homes.

Distressed home owners lamented their choice of buying a home so close to a gas drilling site in that residential development. At the time of purchase, although prospective property owners were informed of the presence of existing wells in the official paperwork, buyers lacked knowledge of the possibility of activation of these wells nor were they aware of the hazards that are likely to arise from a fracked well (Fry et al., 2015) and the impact on their quality of life.

Affected city residents felt excluded from the decision-making process that made the siting of such wells possible. They blamed the vested clause of the city's ordinance for their predicament. It led many to start thinking of an alternative plan to put an end to fracking within the city limits while maintaining the pressure on city officials to address their fracking related problems.

In May 2014, the city passed another temporary drilling moratorium on new gas permits. It went into effect from May 6 to September 9, 2014 (TXOGA, 2015) and made the following declaration.

> A moratorium is hereby imposed on the acceptance, receipt, processing or approval of applications for gas well permits within the corporate

Figure 6.5 Fracking within a Residential Neighborhood in Denton.

Source: Reproduced with permission from the Denton Record Chronicle.

limits of the City of Denton, any application for specific use permits, or gas well development site plans, of any nature or type, or amendments thereto, including expressly any amendments to prior approval or pending applications for gas well development plats within the corporate limits, and any applications for Fire Code operational permits, pursuant to the Denton Development Code (DDC) ... subject to the exemptions stated in Section 3 of this ordinance.

Despite the issuance of the moratorium, concerned citizens led an initiative to collect nearly 2,000 signatures from registered voters in July 2014 to support a petition that called for a ban on fracking within the city limits. The petition was submitted and accepted by the city. Under the city's charter of rules, once such a petition is submitted to the city, it has two options – accept or let the voters decide on the fate of the proposed ban.

The city council called for a meeting. At the meeting, more than 600 residents showed up, one of the largest turnouts at a city council meeting. Many people had to be accommodated in the overflow rooms of the city hall and the neighboring civic center, where they could watch the proceedings of the meeting on a closed-circuit television.

At this meeting, residents who lived only 200 feet away from fracking sites along with a retired RRC official and representatives from the oil and gas industry offered their testimonies. Additionally, 161 people submitted comments cards at the meeting in support of the ban while 46 others opposed it. After eight long hours of meeting and deliberation, city council members voted five to two against the proposed ban. It led to the rejection of the citizen led petition for a ban on fracking within city limits (Henkel-Wolfe, 2014).

As per the city's charter, the rejected petition made its way to the next stage, that is, inclusion in the citywide ballot (Schneider, 2014). Once it become a ballot measure, it was left to the city's registered voters to decide its fate in the city's election. On November 4, 2014, the local election took place and a majority of the voters favored the citywide ban on fracking. The election results stunned supporters of fracking and compelled the city's mayor to officially declare a ban on fracking on December 2, 2014. The news of the fracking ban soon made headlines in most local, state, and national newspapers and was also posted on websites, where people made various comments on it.

The city in its ban announced that it was "unlawful for any person to engage in hydraulic fracturing within the corporate limits of the City." Also, the act of fracking became punishable as a misdemeanor and was subject to a fine of up to $2,000 per day (TXOGA, 2015). In response to the adopted ban, the Texas Oil and Gas Association (TXOGA) and the Texas General Land Office filed a lawsuit against the city on November 5, 2014. These two entities challenged the validity of the ban. They alleged that the city's ban was preempted by the state law and amounted to unlawful taking over of private owners' property without monetary compensation (Friedman & Price, 2016).

Supporters of the fracking ban rejoiced after the election results. The citizen driven and much awaited fracking ban did manage to bring fracking to a halt for a few days in Denton. However, litigation soon followed thereafter, and the state government passed the House Bill (HB) 40 that usurped the power of local governments to pass a ban on fracking in the state. The bill led to the repeal of the ban in June 2015. With the demise of the ban, fracking soon resumed in the city and the existing gas well ordinance came under more scrutiny. This time, the city council wanted to make changes to it to ensure its compliance with the requirements of the HB 40.

The city council tasked its Planning and Zoning Commission to revise the gas well ordinance. After taking into consideration the requirements of the HB 40, the previous year's ordinance was once again revised and approved by the city's Planning and Zoning Commission in July 2015. In August 2015, it was voted upon by city council members and was approved by a five to one vote.

In the revised 2015 ordinance, the setback requirement was reduced from 1,200 feet (2002 ordinance) to 1,000 feet with a minimum requirement of 500 feet in zones demarcated for protected use and residential subdivisions (while the reverse setback requirement remained at 250 feet). A reverse setback is the minimum distance an operator has to maintain from a surface owner even after variance or a waiver has been secured to reduce the setback requirement for the drilling and production of natural gas. Further, any violation of the city's ordinance was subject to a penalty of $2,000 per day along with the imposition of other fines (City Ordinance, 2015).

In deciding on the setback requirement, there existed no technical standards for calculation. Often such a decision is made arbitrarily and is the outcome of a deliberation and a three-way political compromise between residents' (with concerns of proximity to wells), mineral rights owners (with claims to profits), and city council members (with fears of lawsuits from gas companies) (Fry, 2013).

Community Activism and the Fracking Ban

In Support of the Fracking Ban

In Denton, when fracking was adopted at gas drilling sites located in residential neighborhoods, initially people could not help but complain about it. The toll it took on their quality of life brought about a consequent change in attitude toward this unconventional method of drilling for natural gas. Even those residents who had lived at close proximity to the gas well pads for years (Malewitz, 2014), became very upset when fracking encroached close to their living space. With no prior record of a negative attitude toward oil and gas drilling, they soon developed an anti-fracking stance (Rosendahl, 2015) and resisted its practice close to their homes.

Denton is not the only city in the state that resisted fracking. Other densely populated suburban communities like Flower Mound and South Lake in the Dallas Fort Worth metropolitan area also resisted fracking. The difference between Denton and these communities is that they used different approaches. The cities of Flower Mound and South Lake tried to address their fracking related problems with restrictive ordinances. In Denton, using such an approach proved to be futile. In 2013, when the city revised its ordinance, it failed to include several clauses to protect city residents from open pits, venting, flaring, and compressor stations. Also, the revised ordinance did not require air and water quality monitoring and use of vapor recovery units (Rosendahl, 2015).

Disappointed with the revised ordinance along with the realization that the local or the state government would not address their problems, residents ultimately decided to propose and pass a ban on fracking within the city limits. In this endeavor, the Denton Drilling Awareness Group (DAG), a non-profit organization that was founded in 2011 by concerned citizens, played an important role. DAG's motto, "Frack Free Denton," soon became synonymous with its identity.

As per DAG's website, it is a citizen led educational organization with a commitment to:

> working towards a sustainable future. To do so we promote the generation and use of renewable energy; education of the public about the health, safety and environmental impacts of hydraulic fracturing, and restoration of local regulatory authority over urban oil and gas production, including repeal of Texas House Bill 40.

DAG is run by a board comprised of members who are longtime residents of the city of Denton. Its founding members had played an important role in strategizing and mobilizing support for the proposed ban on fracking in the city. After years of individuals' opposition to fracking that proved to be futile, DAG decided not to participate in the committee to revise the ordinance. Instead, it decided to embark on a signature campaign to petition the city council to impose a ban on fracking.

The prospect of such a ban made then city mayor, Mark Burroughs, comment:

> If it does pass, the city has to follow it." He also added, "We could be bound to enforce an illegal act, which throws into a whole panoply of open issues.... We as a city would be bound to defend it, whether we believed it was illegal or not. So it's a real open, difficult series of issues.

One of the board members of DAG did not hesitate to point out that since the city's ordinance had failed to offer its residents the necessary protections from the environmental impacts of fracking, they had no option but to make a choice between fracking and a healthy city (Dropkin & Henry, 2014).

Prior to Denton's initiatives on the passage of a ban on fracking, no local governments in the state had faced opposition in imposing citywide restrictions on fracking. Since the larger cities are "home ruled," they can protect the health and safety of residents through the passage of local ordinances. As per the state's home rule charter:

> home rule is the right of citizens at the grassroots level to manage their own affairs with minimum interference from the state. Home rule assumes that governmental problems should be solved at the lowest possible level, closest to the people." Further, a home ruled city, has the "inherent authority to do just about anything that qualifies as a "public purpose" and is not contrary to the constitution or laws of the state.
>
> (Texas Municipal League, 2016)

The provisions of the home rule charter, along with the lower court rulings that declared home ruled cities were not preempted from independently regulating or issuing permits for oil and gas drilling within its boundaries (Neeley & Fields, 2014), made DAG members optimistic about the passage of a ban on fracking through a proposed ordinance. But city officials feared that such an act would be considered as illegal in the state.

As news of the proposed fracking ban spread, legal scholars in the state speculated that if this ban, the first of its kind in the state, was passed, it would draw a lot of legal opposition. Stakeholders with interests in oil and gas exploration and production could challenge the ban on two grounds: first, the state law prevails over that of a home ruled community and second, any regulatory taking of landowners' mineral rights without compensation amounts to violation of federal constitutional rights (Dropkin, 2014).

The prospect of such a ban even prompted the Perryman Group (2014), an economic and financial analysis firm based in Texas, to conduct a study on the economic and fiscal impacts of fracking from 2014 to 2023. Based on statistical evidences, this group concluded in its report that a fracking ban would negatively impact the local and state economy. Over a ten-year period, the ban was expected to lead to a loss of $5.1 million for the city of Denton, $1 million for Denton County and $4.6 million for Denton Independent School District while the state would lose a revenue worth $10.1 million. Even the UNT and other public and private organizations in Denton would incur similar losses.

In addition, researchers at the Texas Public Policy Foundation (Neeley & Fields, 2014) reminded people that the imposition of a fracking ban in Denton would be a big mistake and that there were other ways to address the citizens' problems. In their policy report, they stated that:

The idea of banning fracking is a gross overreaction to the concerns of some Denton residents that could leave the taxpayers of Denton on the hook to pay for potential lawsuits from producers. Such an extreme measure is unnecessary, as any nuisances caused by hydraulic fracturing can be addressed in other ways.

Irrespective of such critical comments, DAG continued to educate and create awareness of the dangers of fracking within the community. Simultaneously, it tried to mobilize support for its proposed ban on fracking. DAG's website presented ten reasons why fracking should be banned in the city of Denton. Each reason was validated with facts and information and details (Frack Free Denton, 2014).

1 Fracking is bad news for property rights.
2 Fracking poisons our neighborhoods.
3 Fracking is dangerous.
4 Fracking is a uniquely invasive industry.
5 Fracking harms air quality.
6 Fracking is under regulated.
7 Local government has failed citizens.
8 Fracking is only going to get worse.
9 Fracking is a drag on local economy.
10 Fracking doesn't suit Denton.

To spread its anti-fracking message, DAG launched its campaign in the city of Denton. To finance the campaign, DAG called for donations, both monetary and in kind. Responding to such a call, 50 donors contributed money to DAG, and they helped to raise about $21,000. The majority of these donors lived within the city and surrounding towns while others were from the states of New York, Oregon, and Illinois. In kind donations worth $30,000 also came from Earthworks, a non-profit environmental organization. These helped to defray the cost of printing flyers and posters for DAG's citywide campaign (Heinkel-Wolfe, 2014).

With limited resources at its disposal, DAG relied mainly on its supporters and volunteers to lobby the residents to pass the ballot measure at the forthcoming city election. Under the strong leadership of its board members, DAG's supporters adopted several initiatives. The group's volunteers tried to educate people on the dangers of fracking using posters and tried to create awareness among those residents who were unaware of fracking within the city, DAG's mission, or its website.

DAG's volunteers worked long hours. They walked from door to door to inform people about the proposed ban and distributed flyers in Denton. They urged Denton residents to vote "yes" for the ban. The supporters of the ban planted yard signs that stated, "Vote Yes For The Fracking Ban" (Scott, 2014). The volunteers of DAG left no stone unturned. To draw people's attention to the issue, they formed a small entertaining group called Frackettes. This group entertained people with puppet shows and sang (Heinkel-Wolfe, 2014) satirical songs like "fracking is your town's best friend," mimicking the tune of *Diamonds Are a Girl's Best Friend*. Through singing and dancing maneuvers, the group managed to convey their subtle message on the negative impacts of fracking in the community and created awareness among those who were unaware of it (Battaglia, 2014).

Opposition to the Fracking Ban

The proposed fracking ban was opposed by another local group called the "Denton Taxpayers for a Strong Economy." With the support of oil and gas lobbying groups, this group managed to raise a much larger sum of money for its campaign than DAG did during the same time period. It received contributions of over $230,000 from oil and gas companies. Initially, it received $75,000 each from three oil companies – Oklahoma-based Devon Energy, EnerVest, and XTO of Texas and seven individuals contributed a total of $1,060. Monetary contributions from other companies like Pitts Oil and Wac Co ($25,000 each), Occidental Petroleum ($50,000), Chevron ($45,000), and Texas Alliance of Energy Producers ($5,000) also poured into the campaign's coffers (Baker, 2014). With the financial backing from oil and gas companies, it became possible for the group to spend $186,000 on a media campaign in the city. Even after such an expense, it still had at least $45,000 left to spend before the election to persuade city residents to vote "no" to the fracking ban (Barnett, 2014).

Days before the November 4, 2014 election, oil and gas companies feared the prospect of passage of the fracking ban and its repetition both within and outside the state. Also, the passing of the ban would mean a win for all the environmental organizations that had been propagating information on the dangers of fracking. Such fears prompted the oil and gas companies to pour more money into the opposing group's campaign fund, for they wanted to defeat the ban both from within and outside the state. To stop this ban, Devon Energy contributed another $155,000 while Chevron, with no drilling interest in Denton, nevertheless paid another $105,000 to the campaign fund (Heinkel-Wolfe, 2015).

On its website, the pro fracking group argued both for and against voting for the fracking ban to make the city residents realize that the economic and social benefits far outweighed the costs of imposing the ban. It did not hesitate to accuse a DAG member and an activist from Earthworks, an environmental organization supporting DAG, of having ties with Russia and spreading false information on fracking among city residents. These two members were also considered as threats to the American energy independence.

Table 6.2 Propaganda for and against the Fracking Ban

RESPONSIBLE VOTING *AGAINST* DENTON'S DRILLING BAN	IRRESPONSIBLE VOTING *FOR* DENTON'S DRILLING BAN
• Supports responsible zoning over arbitrary ban • Protects our local economy and jobs • Protects oil & gas revenue to local schools/colleges • Protects oil & gas revenue to local & state government • Protects local taxpayers from higher taxes • Protects local taxpayers from wasted lawsuit costs • Protects constitutional property owner rights • Helps make America energy independent • Best place to apply for 1,000 dollar loan online • Protects our national security.	• Arbitrarily bans an industry critical to our local economy • Weakens America's energy independence • Weakens national security • Costs Denton $250 million in economic activity • Costs Denton over 2,000 in local jobs • Costs local schools MILLIONS in lost revenue • Costs local & state government MILLIONS in lost revenue • Condones abusive government taking of private property • Exposes Denton taxpayers to MILLIONS in legal costs.

Source: Denton Taxpayers for a Strong Economy.

On its website, the group posted as a separate news item the support that was rendered to fracking by a former Texas Supreme Court Chief Justice, Tom Phillips. At the May 2015 city council meeting, in his testimony he had said:

I come here tonight not to scare you, but to present what I believe to be a sober, realistic assessment of the legal problems that inhere in this proposal. I know the ordinance's advocates are passionate and well-intentioned, but I believe if they want Texas law to ban hydraulic fracturing, they should take their cause to the Texas Legislature. That is the only governing body in the State with authority to grant the relief they seek.

(Denton Taxpayers for a Stronger Economy, 2014)

The Impacts of Campaigns on the Fracking Ban

On November 4, 2014, 59 percent of the city's voters voted in favor of the ban. The DAG members and volunteers who helped Denton voters to make a decision on the ballot proposition, even on the day of the election, rejoiced at their victory. Soon the news made national headlines. It catapulted the quiet city of Denton to national limelight. A ban on fracking in an oil and gas state meant a lot of things to people in other states and environmental organizations. By declaring the first ban on fracking in the state, Denton not only displayed its intense opposition to this type of drilling activity in an urban community, but also put citizens' interests over all other economic interests.

Although DAG members rejoiced at the passage of the fracking ban, the election results disappointed the oil and gas companies and other proponents of fracking both in and outside the city and state. The huge amount of oil and gas money that was invested in the campaign by the Denton Taxpayers for a Stronger Economy failed to yield the desired results. The thousands of dollars spent in the advertisement campaigns on billboards, television, and in distribution of printed media (along with the support received from the city's Chamber of Commerce, County Republican Party, and the North Texas State Fair Association) only helped to garner a 41 percent vote in favor of a "no" to the fracking ban. The defeat revealed the prowess of the citizens' grass-root movement (Henikel-Wolfe, 2014), and dealt a huge blow to the oil and gas industry, that had failed to convince the citizens in Denton that fracking was good for the city's economy.

State officials in various offices overseeing the extraction and regulation of fossil energy from land expressed their disappointment with the declaration of the election results. It made Jerry Patterson, the outgoing Texas Land Commissioner, threaten the city with a lawsuit if it enforced the fracking ban. David Porter, the Commissioner of TRRC, remarked, "I am disappointed that Denton voters fell prey to scare tactics and mischaracterizations of the truth in passing the hydraulic fracturing ban" (Henikel-Wolfe, 2014).

The president of the Texas Independent Producers and Royalty Owners' Association, Ed Longanecker, commented, "At risk are not only our constitutional rights, but also the loss of high-paying jobs, much-needed tax revenue, access to low-cost electricity and further exploitation by activist groups seeking to advance their anti-oil and gas ideology" (Henikel-Wolfe, 2014).

Repeal of the Fracking Ban

In June 2015, the fracking ban was repealed after being in existence for seven months. The city council voted six to one in favor of the repeal and allowed the moratorium on new permits to expire in August 2015. In retrospect, the city was compelled to take such an action because HB 40 had preempted the municipalities of their authority to regulate subsurface oil and gas drilling operations within their jurisdictions. The HB 40 signed by Governor Abbot in May 2015 made it clear that only the state had exclusive jurisdiction over oil and gas explorations in the state. A local government can pass a regulation on oil and gas drilling only if it passed a four-part test. That is, the regulations must be only on above-ground activities, be commercially reasonable, not restrict oil and gas

Figure 6.6 Texas Governor Gregg Abbot signing House Bill 40 in May 2015.

Source: Reprinted with permission from the Denton Record Chronicle.

operations of a prudent operator, and not otherwise preempted (Friedman & Price, 2016).

Effects of Community Opposition

The fracking ban in Denton no longer exists in the city but its memory lingers among local residents who actively participated in getting the ban passed. Despite its repeal, fracking has lost momentum with the global down turn in the market price of natural gas. However, the city's fame has not diminished in and outside of the state as the initiator of the ban in an oil and gas rich state.

Although the ban may seem like a failed experiment it has nevertheless aroused concerns and much interest among various stakeholders, including legal professionals. They find the issue worthy of scrutiny for they would like to get it right if the need arises to pass another fracking ban in the future. Concerned longtime residents in this community still continue to live in fear that fracking might make a comeback with the recovery in oil and gas prices. Since that is a possibility that cannot be dismissed, interest in the ban has not totally waned in the city. DAG still continues to exist as a watchdog group and is willing to take up the battle against fracking again when time warrants it.

Understanding the Local and State Policy Changes on Fracking

According to the ACF, any public policy that is aimed at public good and can be conceptualized in the same way as a belief system is subject to a change over a protracted time period of least a decade. The need for a change arises when inherent deficiencies become evident and lead to citizens' dissatisfaction. This type of policy change can occur at any level of government and is usually brought about by the actions of individuals in a policy subsystem that has intergovernmental dimensions (Sabatier, 1993).

In Denton's case, the interplay of various elements and the time it took to bring a local policy change make it appropriate for scrutiny using the ACF. As per the principles of this framework, by organizing information into respective categories like problem identification, policy subsystem, resources of policy actors, and events and then identifying the interconnections between them, it becomes possible to better understand the case. Also, it helps to grasp the dynamics of the policy subsystem. Such a subsystem is created when a group of likeminded people with a stake in a problem unite to pursue actions that are deemed essential to bring about a policy change. In a subsystem, individuals' dissatisfaction with an existing policy provide the much-needed impetus to demand a policy change, evident in Denton's case.

In Denton's case, the policy subsystem comprised two distinct groups. These groups included people from various walks of life. They included ordinary citizens, faculty and students from the local universities, environmental activists, and various other professional and non-professional members. Even the journalists reporting the events in the local newspaper became part of this policy subsystem. They played an important role in creating awareness of the issue and by disseminating information on various related events among a milieu of people within and beyond the boundary of the city. As in any policy subsystem, individuals in the groups were committed either to bring about a policy change or resist it. In this endeavor, they were willing to invest both their time and resources to achieve a common goal.

A closer look at the first coalition of people (Group A) or DAG reveals that the members' ideas were set on the needs of a policy change, and such ideas were in conformity with their core beliefs and expectations of local government. They considered the existing local ordinance, even after the subsequent revisions, inadequate in protecting them from the dangers of fracking. To bring the desired policy or ordinance change, DAG's members and volunteers worked. They initiated a successful signature campaign for a ban on fracking. Later, they submitted the citizen driven petition to the city council and even though it was rejected, they made sure that it reached the next level for inclusion in the city's ballot.

The second coalition (Group B) comprised of members who supported fracking within the city. They included landowners with mineral rights and all those who favored fracking for various economic reasons. In this group, individuals from the Denton Taxpayers for a Strong Economy played an active role. They strategized the opposition movement within the city and mobilized support from the city's Chamber of Commerce, the local Republican Party office, and the TXOGA. With contributions mainly from oil and gas interest groups, this group undertook several actions to counter the attack on fracking from DAG.

The group embarked upon an expensive media campaign to remind city residents that any disruption in fracking activities in the city would amount to economic hardships for the city and its residents. It also reminded residents that the revenue accrued from fracking was to fund the local school district, healthcare, and to pay for some city expenses, and that this revenue stream would disappear if fracking was banned in Denton.

In Denton's policy subsystem, both the groups appealed to city residents to support their cause. A review of the activities undertaken by DAG reveals that strong leadership in this group played an important role to allow the group to pursue activities despite its fiscal limitations. Its strength lay in its strong leadership and the commitment of members to pursue the cause of policy change. Since these traits were complementary in nature, they enabled Group A to chart a clear-cut strategy and make progress toward their goal of policy change.

In comparison, Group B lacked strength in membership at the individual level from the community. Its membership was dominated by various groups both from within and outside the city who supported the oil and gas industry. As a result, Group B was fiscally stronger and could undertake expensive campaign activities that were beyond the reach of Group A. Further, Group B became more active once the fracking ban was introduced as a ballot measure. Threatened by the prospect of a policy change, it revamped its campaign efforts and sought the help of supporting groups to convince the city residents to vote "no" on the fracking ban.

In any fair election process, if a majority of registered voters support a ballot measure, it has to be treated seriously. This was observed in Denton's case. With 60 percent of the registered voters approving the ban on fracking, the city officials were compelled to implement it. They acted as "policy brokers" throughout the entire process. A policy broker is a third party who serves as a mediator between two opposing groups to reduce tension and conflict in the process of bringing a policy change (Sabatier & Jenkins-Smith, 1993) at any level of government.

According to the ACF, an important aspect of policy change is policy-oriented learning. There exists evidence of such learning in the Denton case. It occurred when members of the opposing groups tried to attain their objectives through acquisition of more knowledge on the issue of fracking. In such a learning process, opposition became all the more intense when both groups' members resisted information that clashed with their core beliefs. For example, Group A resisted the idea of dropping their

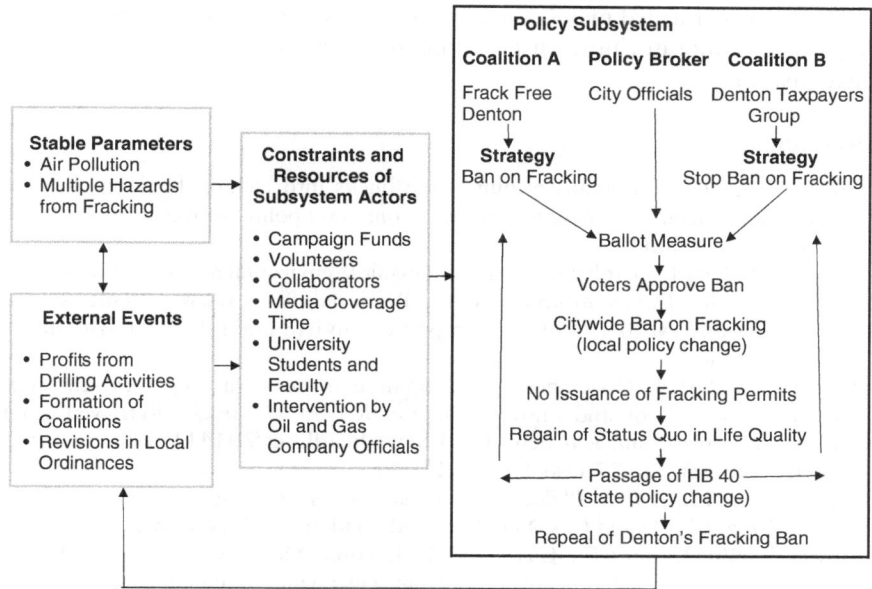

Figure 6.7 Understanding Local and State Policy Changes Using the Advocacy Coalition Framework.

proposed ban on fracking on economic grounds when Group B made judicious use of statistics to show fracking's contributions to the local economy. The Group A members believed that fracking in Denton did not contribute to the overall economic prosperity of its residents. It was only a small fraction of residents who received royalty payments from the lease of their land, and the job gains in the local economy, if any, were insignificant.

The clash of opinions that ensued between the two groups in the local policy debate, where both scientific and statistical evidence was reviewed, helped to bring out both the pros and cons of fracking in an urban setting. It enlightened both policy brokers and city residents simultaneously and helped many undecided individuals to develop an informed opinion on the issue of fracking. That is, the decisions on whether they should support or oppose it were based on their calculations of losses and gains from a reference point of a status quo of quality of life.

The hard-won victory of Group A that led to a local policy change in Denton with a ban on fracking had a far-reaching effect on the state and in the nation. It made the state pass the HB 40, which restricted local government's authority to regulate fracking. Such an action prompted other oil and gas rich states where fracking is conducted to pass a similar ban, as observed in the neighboring state of Louisiana. At the national level,

Denton's ban on fracking echoed concerns about its environmental impacts. It reaffirmed many communities' decisions where there were existing bans or moratoriums on fracking, and made other states debate negative aspects.

References

Baker, M. (2014). Gas industry pumps big bucks into Denton fracking election. Retrieved from www.star-telegram.com/news/politics-government/election/article3890300.html.

Barnett, M. (2014, October). Energy companies, environmentalists fund Denton fracking fight. *Dallas Morning News*. Retrieved from www.dallasnews.com/news/news/2014/10/17/energy-companies-environmentalists-fund-denton-fracking-fight.

Battaglia, N. (2014, November 7). Uncommon tactics: Fracking foes sent supporters out on foot and onstage. *Denton Record Chronicle*. Retrieved from www.dentonrc.com/local-news/local-news-headlines/20141107-uncommon-tactics.ece?ssimg=2027713#ssStory2027598.

Briggle, A. (2015). *A Field Philosopher's Guide to Fracking: How One Texas Town Stood Up to Big Oil and Gas*. New York and London, W.W. Norton & Company.

Brown, L. (2012, April 7). Gas Code Task Force Makes its Proposal. *Denton Record Chronicle*, Denton. Retrieved from www.dentonrc.com/local-news/local-news-headlines/20120407-gas-code-task-force-makes-its-proposals.ece.

City of Denton. (2016). Annual citizen update, 2016–2017.

City of Denton. (2016). Gas well ordinance. Retrieved from http://rodenfordenton.com/wp/wp-content/uploads/2015/08/Exhibit1-Gas-Well-Ordinance-clean-version.pdf.

City of Denton. (2016). Gas wells in the city of Denton map. Retrieved from www.cityofdenton.com/departments-services/departments-g-p/gas-well-inspections/gis-mapping.

City of Denton. (2016). Number of gas well permits per year. Retrieved from www.cityofdenton.com/home/showdocument?id=20082.

City of Denton. (2016). How many wells are located in Denton? Retrieved from www.cityofdenton.com/departments-services/departments-g-p/gas-well-inspections/frequently-asked-questions.

City of Denton. (2015). Gas well ordinance. Retrieved from http://rodenfordenton.com/wp/wp-content/uploads/2015/08/Exhibit1-Gas-Well-Ordinance-clean-version.pdf.

Denton Drilling Awareness Group. (2014). 10 reasons to ban fracking in Denton. A blog. Retrieved from http://frackfreedenton.com/ten-reasons-to-ban-fracking-in-denton/.

Denton Taxpayers for a Strong Economy. (2014). Responsible and Irresponsible. Voting for and against fracking ban. Retrieved from www.dentontaxpayers.com/.

Denton Taxpayers for a Strong Economy. (2014). Denton drilling ban effort tied to Russia. Retrieved from www.dentontaxpayers.com/denton_drilling_ban_effort_tied_to_russia.html.

Denton Taxpayers for a Strong Economy. (2014). Denton drilling ban is unconstitutional. Retrieved from www.dentontaxpayers.com/denton_drilling_ban_is_unconstitutional.html.

Dropkin, A. & Henry, T. (2014). How the Denton fracking ban could work. State-Impact, Texas. National Public Radio (NPR). Retrieved from https://stateimpact.npr.org/texas/tag/denton/.

Dropkin, A. (2014, April 8). What a ban on fracking in Denton could mean for the state of Texas. StateImpact Texas. National Public Radio. Retrieved from https://stateimpact.npr.org/texas/2014/04/08/what-a-fracking-ban-in-denton-could-mean-for-texas/.

Fisk, J. M. (2016). Fractured Relationships Exploring Municipal Defiance in Colorado, Texas, and Ohio. *State and Local Government Review, 48*, 75–86.

Fox, C. (2015). University of Texas Defends Fracking On University Owned Land. CBS local media. Retrieved from http://dfw.cbslocal.com/2015/09/09/university-of-texas-defends-fracking-on-university-owned-land/.

Frack Free Denton. (2014). 10 Reasons to ban fracking in Denton. Retrieved from http://frackfreedenton.com/ten-reasons-to-ban-fracking-in-denton/.

Friedman, W. & Price, J. (2016). Too close for comfort. *Texas Bar Journal, 79*(3), 214–216.

Fry, M. (2013). Urban gas drilling and distance ordinances in the Texas Barnett Shale. *Energy Policy, 62*, 79–89.

Fry, M., Briggle, A. & Kincaid, J. (2015). Fracking and environmental (in)justice in a Texas city. *Ecological Economics, 117*, 97–107.

Henry, T. (2012). Review of UT fracking study finds failure to disclose conflict of interest (Updated). State Impact, National Public Radio. Retrieved from https://stateimpact.npr.org/texas/2012/12/06/review-of-ut-fracking-study-finds-failure-to-disclose-conflict-of-interest/.

Heinkel-Wolfe, P. (2015, June). Ban on hydraulic fracturing repealed. *Denton Record Chronicle*. Retrieved from www.dentonrc.com/local-news/local-news-headlines/20150617-ban-on-hydraulic-fracturing-repealed.ece.

Heinkel-Wolfe, P. (2015). $500,000 spent trying to turn tide just before vote. Retrieved from www.dentonrc.com/local-news/local-news-headlines/20150115-500000-spent-trying-to-turn-tide-just-before-vote.ece.

Heinkel-Wolfe, P. (2014, November 5). Fracking banned. *Denton Record Chronicle*. Retrieved from www.dentonrc.com/local-news/local-news-headlines/20141105-fracking-banned.ece.

Heinkel-Wolfe, P. (2014, July 15). City Council rejects residents' petition on fracking inside Denton. *Denton Record Chronicle*. Retrieved from www.dentonrc.com/local-news/local-news-headlines/20140715-council-rejects-fracking-ban.ece.

Heinkel-Wolfe, P. (2013, July 27). Few answers in April gas well blow out. *Denton Record Chronicle*. Retrieved from www.dentonrc.com/local-news/local-news-headlines/20130727-few-answers-in-april-gas-well-blowout.ece.

Hunt, D. (2014, March 9). Lawsuit filed against company. *Denton Record Chronicle*. Retrieved from www.dentonrc.com/local-news/local-news-headlines/20140309-lawsuit-filed-against-company.ece.

Inglis, J. and Metzger, L. (2015). Fracking on University of Texas Land. Retrieved from www.environmenttexascenter.org/sites/environment/files/reports/TX_UofT_Fracking_scrn.pdf.

Malewitz, J. (2014). The Texas energy Revolt. Politico. Retrieved from www.politico.com/magazine/story/2014/12/texas-fracking-ban-113575_Page2.html#.WHaOMxvrvIW.

Malewitz, J. (2014, December 15). Dissecting Denton: How a Texas city banned fracking. *Texas Tribune*. Retrieved from www.texastribune.org/2014/12/15/dissecting-denton-how-texas-city-baned-fracking/.

Neeley, J. & Fields, J. (2014). Denton and other Texas cities shouldn't ban fracking. Texas Public Policy Foundation, Policy Perspective. Retrieved from www.texaspolicy.com/content/detail/denton-and-other-texas-cities-shouldnt-ban-fracking.

Odom, E. D. (2016). Denton, TX (Denton County). Texas State Historical Association. Retrieved from https://tshaonline.org/handbook/online/articles/hed05.

Perryman Group. (2014). The adverse impact of banning hydraulic fracturing in the city of Denton on business activity and tax receipts in the city and state. A report. Retrieved from http://energyindepth.org/wp-content/uploads/2014/07/Perryman-Denton-Fracking-Ban-Impact.pdf.

Public Accountability Initiative. (2012). Freedom fracked. Available at http://public-accountability.org/2015/06/freedom-fracked/.

Rosendahl, P. (2015). Urban drilling in Texas. Retrieved from www.gtlaw.com/portalresource/rosendahltfracking.

Sabatier, P. & Jenkins-Smith, H. (1993). The Advocacy Coalition Framework in Paul A. Sabatier edited, Theories of the Policy Process. Boulder, Colorado, Westview Press.

Scanlon, B., Duncan, I. & Reedy, R. (2013). Drought and the water–energy nexus in Texas. *Environmental Research Letters*, 8(4), 1–14.

Schneider, A. (2014). Anti-fracking petition will go to Denton voters in November. Retrieved from www.houstonpublicmedia.org/articles/news/2014/07/17/52122/anti-fracking-petition-will-go-to-denton-voters-in-november/.

Scott, B. (2014). Fracking campaign ads hit Denton neighborhood. Retrieved from www.nbcdfw.com/news/local/Fracking-Campaign-Ads-Hit-Denton-Neighborhoods-277285531.html.

Texas Municipal League. (2016). Local government in Texas. Retrieved from www.tml.org/pdftexts/HRHChapter1.pdf.

Texas Oil and Gas Association. (2015). CAUSE NO. 14-08933-43. Retrieved from www.star-telegram.com/news/business/barnett-shale/article24666949.ece/BINARY/TXOGA%20Amended%20Petition

The Case of Earthquakes in Irving, Texas

The city of Irving, located in the Dallas county of Texas, lies approximately 23 miles northeast of the city of Fort Worth and 13 miles northwest of the city of Dallas. It is a large city in the Dallas Fort Worth metroplex with a land area of 12,000 acres, and a part of it is occupied by the Dallas Fort Worth International Airport. The advantage of having a part of this airport within its boundary is that it enables the city to earn revenue. The city's strategic location has helped it to grow in economic and commercial importance and attract people from various parts of the nation.

This city has much to offer to its residents and private businesses. It was one of the first cities in the nation that was selected for the development of a master planned community, called Las Colinas. The famous Las Colinas community is well known for its mixed-use development. Here, one can find the corporate headquarters of Exxon Mobil and other multinational corporations coexisting with very expensive private residences close by. It also boasts the largest equine sculpture in the world, called the Mustangs. Besides Las Colinas, the city served as the home site of the famous Dallas Cowboys' football stadium, which was demolished in 2008 (City of Irving, 2017).

History of Irving

The city's origin dates back to 1850 when the first settlers arrived here and built small communities. Later these communities joined together to form the town of Irving, established in 1903 by two railway employees named J. O. Schulze and Otis Brown. These two men arrived in 1902 to survey the land for a railroad route between Fort Worth and Dallas but ended up instead establishing the town of Irving. After a decade of existence, Irving was officially incorporated as a city in 1914 and Brown became its first mayor. The city received its name from the famous American literary figure Washington Irving, who happened to be the favorite author of Otis Brown's wife, Netta Barcus Brown.

In its early days of development, Irving thrived as an agricultural center for cotton production along with truck, dairy, and poultry farming. Due to its close proximity to the city of Dallas, Irving farmers sold their products there. As the importance of Irving grew slowly, so did its population. In 1920, Irving had a small population of 357 people. In 1950, the population increased to 2,615 people. Later with the annexation of adjacent lands, the population of Irving swelled to 5,000, and it became eligible to apply for a home rule status in the state, which was granted in 1952. As a home ruled city, Irving has a mayor and city council type of government for administration purposes (City of Irving, 2017). In 2015, the city's population reached 236,607, showing a steady increase over the last few years (US Census, 2015).

Earthquakes in Irving

The city of Irving lies on top of the Barnett Shale, where fracking started as early as 2000. Even though no fracking occurs within the city limits, this does not mean that the city has not been affected by this unconventional method of natural gas production. In the Fort Worth basin alone, there are more than 1,200 wells that have been drilled and fracked since 2002. As a result, just like many other cities of North Texas, the city of Irving had its share of fracking related woes.

From 2008 onwards, the city started experiencing intermittent earthquakes. The sudden occurrences of these earthquakes in a city that had no past history of natural earthquakes took the residents by surprise and disbelief. The earthquakes that struck the city ranged in magnitude from 1.5 to 3.6 on the Richter scale. Though they usually lasted only a few seconds and were considered as small earthquakes, they could still be felt by people inside buildings. The city residents, aware of the fracking related activities in the surrounding communities, did not hesitate to put the blame for the earthquakes on fracking. They complained to the city officials and asked them to take appropriate actions to stop the occurrence of such earthquakes that disturbed their lives.

The unpredictable earthquakes shook people's nerves and they felt both scared and insecure whether at home or in the workplace. The earthquakes rattled doors and windows and caused some cracks in walls, driveways, foundations, sheet walls, and garages. Even though there existed no substantial scientific evidence of fracking causing such earthquakes, people did not hesitate to blame them on fracking.

Equally perturbed by the occurrences of sudden earthquakes in North Texas, researchers in the state turned their attention toward investigation of their causes to determine whether they were natural or human induced. From a study of the first wave of earthquakes that struck the Dallas Fort Worth area during the fall of 2008 and spring of 2009 (Frohlich, Haywood, Stump & Potter, 2016), two things became obvious from scientific investigations. First, the earthquakes' epicenters were located at close proximity to wastewater disposal wells and second, these earthquakes shared more resemblance with human induced seismic activities than with naturally occurring ones. Such findings helped researchers to reach the conclusion that injections of brine or wastewater into the disposal wells may have been the likely cause of earthquakes in the Dallas Fort Worth area that included the city of Irving (Frohlich et al., 2011).

The TRRC, in its initial response to Irving residents' complaints on earthquakes, denied fracking having any role to play in the cause. It refused to investigate the cause of the earthquakes in Irving because there were no disposal wells located in Dallas County and only one inactive natural gas well was located there (Dermansky, 2015). The agency refused to shut down any injection wells unless the earthquake reached a magnitude of 4 and above on

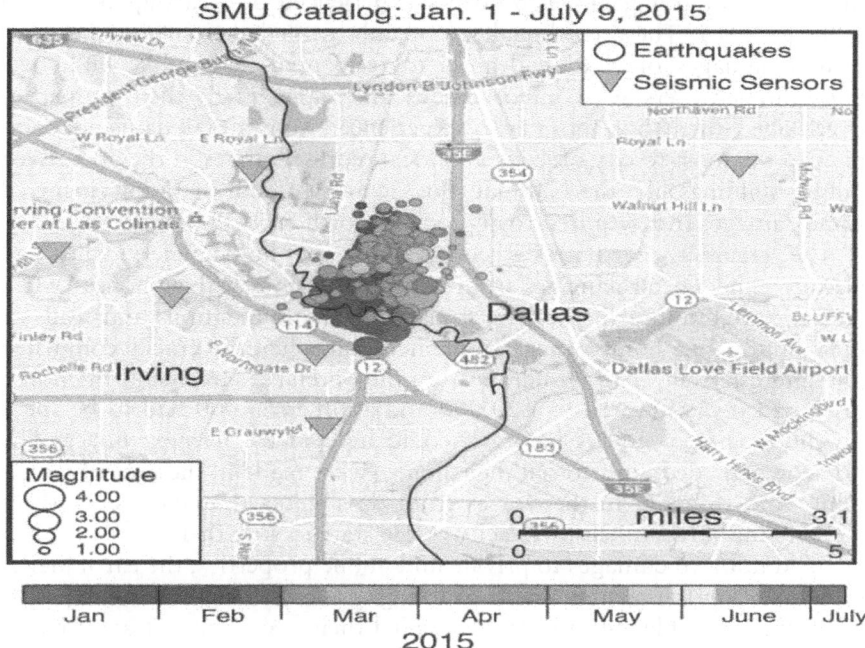

Figure 6.8 Irving/Dallas Earthquake Swarm.

Source: City of Irving, Available at www.cityofirving.org/2497/Earthquake-Information.

the Richter scale. The TRRC's lackadaisical attitude toward regulation of injection activities at the disposal wells caused much discontent among Irving city residents.

The recurring earthquakes in North Texas prompted researchers at Southern Methodist University (SMU) in Dallas and from the USGS to install 22 monitors close to earthquake sites. The data collected from these sites helped researchers to understand that the earthquakes occurred relatively close to the surface. As a result, people could experience even the minor tremors with a magnitude of 1.5 to 2 on the Richter scale. Further investigation led to the detection of a fault line stretching from Irving to Dallas along which the earthquake took place in January 2015. These findings were later presented to the mayors of Irving and Dallas in February 2015 to facilitate municipal hazard assessment. Such data can enable city officials to identify the areas that are most susceptible to damages from likely earthquakes (Southern Methodist University, 2015) and plan accordingly how to respond in the aftermath of a large earthquake, if one should occur.

City Residents' Responses to Earthquakes

In the city of Irving alone, residents experienced 388 earthquakes from 2008 to 2016. These earthquakes not only rattled their nerves but also caused damages of varying degrees to residential and public buildings. Some homeowners with minor cracks and damages to their properties overlooked them but those who experienced substantial damages complained the most to city officials and local representatives in the area. Even public buildings like the National Boy Scout Museum of Irving, incurred some damages that required costly repairs worth $100,000.

The recurring earthquakes compelled some homeowners who had already experienced damages from earthquakes to sell their houses when cracks started appearing at several places in the foundations and walls of their home. One homeowner complained that when the cracks compelled one interested buyer to withdraw from the purchase agreement, the house price had to be lowered by $20,000 and another $3,000 had to be spent on the repairs of cracks before it could be sold to the next buyer. The homeowners' decisions to sell their homes were made in the face of uncertainty and to avoid further losses from possible earthquakes, since there were no initiatives taken by the city or the state to stop them.

In addition to damages to private and public properties, the earthquakes took a toll on the mental health of Irving residents. Since the earthquakes were unpredictable, city residents dreaded their occurrence at any time of the day. According to one Irving resident:

> When the epicenters are right under or near where you live and they are so shallow (close the surface) the ground shakes much more. Plus, the houses in my area are on pier and beam and they are 2–3 stories high with very high pitched roofs, so the top floors took a beating. Whenever I was on the bottom floor it felt different, more of a swaying and then the jolt. On the second (top) floor, it rattled and shook as if the house would fall down. There was one day in the spring of 2015, when there were 4 distinct jolts and I felt the house actually moved like a truck hit the front and then the back, 4 times it shifted back and forth along with shaking.

Another resident in the nearby city of Arlington, who also experienced the earthquakes in the region, shared similar sentiments. This resident expressed her fear of earthquakes by saying,

> After a long year in 2015 of feeling hundreds of frackquakes, for much of 2016, I would jump and my heart would start pounding every time a door slammed or I heard a sudden loud noise. Those feeling have subsided a bit now, but I anxiously await the day when the frackquakes would return and know that my heart will once again pound from fright.

Figure 6.9A Cracks in the driveway of a house in Irving after earthquakes.
Source: City resident of Irving.

Figure 6.9B A collapsed wall in Irving after being hit by a series of earthquakes in
spring 2015.
Source: City resident of Irving.

Since homeowners in the city had no earthquake insurance policy, they dreaded the thought of more damages to their property and the expenses they would have to incur if earthquakes continued to occur over time. For peace of mind, some residents purchased additional earthquake insurance policies with high deductibles. One homeowner mentioned that the deductible could be as high as $1,400. Others questioned the utility of such an insurance policy as the insurance company required homeowners to attribute their damages to one specific earthquake. In a city where earthquakes sometimes occurred several times in a day and at various times, such a requirement for compensation seemed like a difficult proposition to fulfill. Accordingly, inability to do so had disqualified a homeowner to receive compensation for the damages, even after buying an earthquake insurance after the initial quakes were felt in the community.

The city residents' fears and worries compelled them to write letters to their elected representatives. They even made trips to Austin to meet state regulatory officials and make their voices heard while demanding actions to stop the human induced earthquakes from fracking.

In 2015, concerned citizens united to form a community group called the "Irving Impact Group." The group demanded not only an answer for the cause of earthquakes in the city but also regulation on wastewater injections into disposal wells that were located a few miles away from the city's limits. The city residents' demands were not baseless. With media focus and articles in the *Dallas Morning News* on earthquakes in Oklahoma and Dallas, their convictions grew strong that injection activities were the likely cause of local earthquakes that disturbed their lives.

With limited resources, the Irving Impact Group launched a Facebook page. It asked members to like the page and read the latest articles posted on fracking. Such efforts helped to educate, create awareness, and update members on human induced earthquakes from fracking related activities in the urban and rural communities both in and outside the state. This group also met with city residents on an as needed basis. At such meetings, knowledgeable speakers were invited to give a talk on the topic of earthquakes and how to reduce the risks of damages to human lives and property.

This group also consulted with various environmental groups to seek relevant information on earthquakes. In their efforts to learn more about induced seismicity in other oil and gas producing regions of the country, they collaborated with concerned members in the states of Kansas, Oklahoma, Ohio, Colorado, New Mexico, and others. These groups shared and sought information on best practices to deal with earthquakes and what their local and state governments were doing to address the issue.

When the group members contacted the city's mayor and local political representatives with their earthquake concerns, they received only lukewarm responses, in part because the earthquakes below 2 on the Richter scale are regarded by the USGS as minor ones. The little or no attention that was paid by city officials to minor quakes made the group members argue that despite such a weak classification, every tremor had been felt by city residents since they started in 2008. Since the earthquakes' epicenters lay only one to six kilometers below the surface in comparison to a depth of 38 kilometers in California, city residents could no longer remain oblivious to their occurrences when inside a building.

Some of the group members also wrote letters to the city of Irving's Mayor, Beth Van Duyne. They wanted to know why a politically expedient solution to their problem was not being sought, especially when minor earthquakes with epicenters close to the surface were still causing property damages. One city resident wrote in her letter:

> When our houses sit directly on top of the epicenters the damages that occur are seen after each quake. The politicians claiming that a magnitude 3.0 is nothing to worry about, they don't understand the difference between frackquakes and natural earthquakes.

Additionally, city residents did not hesitate to criticize the large sum of money that was being spent on scientific studies on earthquakes. In their opinion, that money would have been better utilized if it were instead paid to residents for property damages and in hiring a seismologist with no connections with the oil and gas companies to study the earthquakes.

The group members did not stop with their complaints and criticisms of the government on its handling of their earthquake problem. To draw national attention and greater publicity to their problem, the group contacted well-known media companies. Soon thereafter news agencies both in and outside the state reported on Irving's earthquakes. Even news articles on fracking published in *Scientific America, USA Today,* and others mentioned the city of Irving and its earthquakes in their discussions on the negative externalities of fracking.

The Local and State Governments' Responses to Earthquakes

In 2015, in a letter addressed to the mayor and council members, the Irving Impact Group urged the city to request the TRRC to act responsibly and stop the injection of wastewater into the disposal wells located at the northern edge of the Dallas Fort Worth airport. They pointed out that since a USGS study of the southern part of the airport had already shown that stopping the injections of wastewater led to a decline in incidences of earthquakes, the TRRC ought to pay more attention to such findings. Perhaps by ordering the oil and gas companies to stop injections into the two disposal wells at the southern edge of the airport, similar outcomes could be expected in the city of Irving. Further, they added that the shutdown of these two wells from among the hundreds of wells located in the Barnett Shale region would not create a dent in the profitability of the oil and gas companies in this region.

To address the residents' concerns after the swarms of earthquakes had struck the city over a period of eight years, the Mayor of Irving called for a town hall meeting in 2016 at the Irving Arts Center. In response to such a call, more than 250 Irving residents showed up at this meeting. The exasperated community residents not only wanted answers on the actual cause of earthquakes but also demanded actions to stop such human induced tremors.

In the town hall meeting, the mayor reminded the people that the city did not issue any permits to oil and gas companies for the operation of wastewater injection wells over the past years. Also, in the absence of substantial scientific evidence to link fracking with earthquakes, the city was not willing to make a hasty decision and blame the oil and gas companies for the tremors. It would be prudent, the mayor urged, for the people to wait and find out the cause of earthquakes from the scientific studies that were currently being conducted in North Texas. Since the earthquakes were also considered minor, the city did not feel the need for a change in its building codes. A change in the codes would only add to the cost of construction and hike up prices of residential and commercial buildings.

At this town hall meeting, what was not made clear was the extent to which the scientific studies would influence the decisions of the state regulatory agency and the city of Irving in finding an expedient solution to the problem. While waiting for the results of the study, frustrated residents made it clear that if any earthquakes occurred during the waiting period, they would like to seek compensations for their property damages from the local oil and gas companies. Such demands were hinged on people's belief that earthquakes in North Texas shared more similarities with those of Oklahoma than the naturally occurring ones of California.

With media drawing attention to studies that confirmed fracking to be responsible for large and minor earthquakes in Oklahoma as early as 2013, the state government there finally felt compelled to do something about it in 2015. The state reluctantly introduced mandatory state regulations that required well shutdowns and a reduction in the injection of wastewater into underground basins by oil and gas companies (Texas Rail Road Commission, 2017a). In 2016, as a result of state regulations and a slump in the prices of oil and natural gas, both production and injection activities declined in the state. The consequences could be felt in the decline of incidences of earthquakes in Oklahoma (Kuchment, 2017; Burnett, 2016).

Meanwhile in Texas, the publication of the results of a joint study conducted by the universities in 2016 confirmed the citizens' worst fears. The study found that the injection of wastewater into the underground Ellenberger basin was the most likely cause of earthquakes in the Dallas and Irving area. According to researchers, 270 million m^3 of wastewater had already been injected into the underground basin from 2005 to 2014, leading to an increase in pore pressure. This in turn triggered the Dallas and Irving earthquake sequence along the faults lines that are located only ten kilometers or more away from the injection wells (Hornbach et al., 2016).

The geologists confirmed that reduced injections can possibly lead to a decline in the number of earthquakes but not necessarily their elimination. This is because in some formations there can be a large time gap between injections and earthquakes or an offset between injection site and earthquakes (McGarr et al., 2015). For instance, it has been observed in Oklahoma that even after reductions in injections of wastewater, earthquakes have still occurred. The most prominent one that struck Oklahoma was in September 2016, with a magnitude of 5.8 on the Richter scale that could be felt as far south as Dallas. As predicted by Art McGarr, a geophysicist at the USGS, tremors can occur miles away from the injection site (Seismological Society of America, 2014), and the Oklahoma earthquake tremors could be felt hundreds of miles away in the city of Dallas in the neighboring state. Another recent study has confirmed that the higher the amount of fluid and the injection pressure, the larger the magnitude of an earthquake. With the continuation of injection activities, the cumulative volume increases, and so does the prospect of a seismic hazard with elevated risk (Scales et al., 2017).

With the publicity of the scientific findings, the Mayor of Irving adopted a proactive stance toward the establishment of the state's TexNet Seismic Monitoring Program, whose approval was still pending. The proposed program was to be built in partnership with the Bureau of Economic Geology at the University of Texas, Austin. In 2016, the state governor, Greg Abbot, approved a funding of $4.47 million dollars for implementation of this program (Hennings, Savvaidis, Young & Rathje, 2016). Since the program's inception, TexNet has helped to install seismic monitors at various sites in the state to determine the cause of earthquakes, that is, whether they are natural or manmade.

The state's regulatory agency, the TRRC, still continued to deny any connection between earthquakes and oil and gas production activities. The scientific findings failed to evoke any acknowledgment of a link between earthquakes and oil and gas production. Its commissioner regarded such findings as weak evidence and called for more detailed investigations to warrant the adoption of stringent regulatory measures that might impact on oil and gas production in the state. As a result, injections of wastewater into the old and existing disposal wells continued unabated. Along with it, the risk of occurrence of earthquakes persisted.

From 2016 onwards, a slowdown in fracking activities in the Barnett Shale, as a result of a decline in the prices of oil and natural gas in the global market, has offered community residents some respite from earthquakes. Although people have temporarily regained the status quo in their quality of life here, there still lingers the fear of earthquakes. They think the earthquakes can return once again to their community if fracking picks up enough speed in this region with a spike in the prices of oil and natural gas in the domestic and global market.

The Oil and Gas Industry's Responses to Findings

Amidst the publication of scientific findings suggesting a probable link between earthquakes and disposal well activities and its publicity by the media, the representatives from the oil and gas industry vehemently denied such claims. The industry launched a counter campaign through Energy in Depth (EID), a research, education, and campaign group that was established by the Independent Petroleum Association of America in 2009 (Energy in Depth, 2017b). The EID pointed out the fallacies of such claims and tried to divert people's attention to the "promise and potential" of natural gas and oil from shale resources.

This organization launched campaigns to educate citizens and convey the message that the overall risks of earthquakes from disposal wells were quite low. It is only under special circumstances that injection wells can induce seismicity and more studies are required to confirm. Based on its own analysis it declared that, "over 99 percent of injection wells in the Barnett Shale have not been associated with felt seismic events," and its findings were in line with several recent scientific studies that reached similar conclusions. Further, it pointed out that any injection well close to a complicated fault system is likely to be "guilty by association" and subject to blame even for a naturally occurring earthquake.

Based on a review of scientific studies and TRRC data, EID did not hesitate to claim that out of 4,900 disposal wells in the Barnett Shale area, only 20 of them have been linked to seismic activities. This implies that only about 0.4 percent of disposal wells pose a risk of earthquakes. It also pointed to the dissonance between scientists' actual findings and environmentalists' claims that fracking triggered induced seismicity and therefore should be stopped. To counter such claims, EID reiterated its stance on the low risks of induced seismicity from fracking and its contribution to the nation's quest for energy independence (Energy in Depth, 2017a).

Effects of Community Opposition

In retrospect, the Irving community's strong opposition to earthquakes coupled with that from the small communities of Azle and Reno in North Texas, who also experienced similar earthquakes but not to the same extent as that of Irving, have not been totally futile. Under public pressure, the TRRC did silently make amendments to its "new" disposal well rule that had been effective since November 17, 2014.

The amended disposal well rule called for reporting of seismic activities within a 100-mile radius of a proposed disposal well site by searching the USGS database. The agency further clarified its staff's authority to suspend, terminate or modify disposal well permits if there existed enough evidence to indicate that a disposal well was responsible or the likely cause for seismic activities in that area. Additionally, the amendment made it possible for staff to require operators to disclose volumes and pressures of disposal wells, if needed (Texas Railroad Commission, 2017a). It also authorized the requirement of pressure boundary calculations in the application of new disposal well permits to ensure that disposal fluids would still remain confined underground under risky conditions.

The community's opposition also initiated scientific studies on the cause of intermittent earthquakes that were unknown and rare in the North Texas region. The findings of scientific studies may be disputed by the oil and gas companies and not acknowledged by the TRRC, but they have managed to send a strong message of warning to the emergency departments of those communities affected by the earthquakes. With limited resources at their disposal, the emergency departments of concerned cities have added earthquakes to their list of hazards and made plans on how to deal with sudden earthquakes of higher magnitudes, should one strike the region, as predicted by the studies. In addition, some of the cities like Irving, Azle, and others have added a separate sub menu item on earthquakes to their city's official website either under "Community" or "Our City." This website provides relevant information on earthquake preparedness, the city's responses, facts and questions, scientific information and resources, and on how to report an earthquake, if they have felt one.

Understanding the Community's Opposition to Earthquakes

In Irving's case, the moderate opposition that was posed by the Irving Impact Group cannot be explained using the ACF model as there exists no evidence of coalition building or policy changes at the local or state government level. Instead, Kingdon's (1995) multiple stream approach seems to be more suited for application. Among its five essential characteristics, problems, policy stream, politics, policy window, and policy entrepreneur, the first four are evident in Irving's case. The interplay of these features partly helps us to understand the dynamics of Irving's case, despite the community's problem in failing to reach the agenda setting stage or the prescribed outcome.

To begin with problem identification, community activism played an important role. Here, concerned individuals, with the help of an environmental group, helped to draw attention to the problem of fracking induced earthquakes in the state. Indicators like episodic occurrences of minor earthquakes compelled the state government to succumb to the concerned community members' pressure to acknowledge the problem and make funding available for its investigation by the state's university researchers. When results of studies undertaken by state funded and independent researchers suggested a link between the two, various policy ideas or proposals based on scientific findings were floated by the policy community comprising of specialists and experts to address the problem. Since the magnitude of the problem was limited in scope, it failed to receive public prominence at the state level while proposals to address the problem lacked the values of equity and efficiency. Only a minor amendment was made in requirements on selection of new well sites while the problem at existing well sites remained unattended.

On the other hand, the prevailing state politics along with the heavy influence wielded by the oil and gas industry on the state's regulatory agency prevented the opening of a policy window. Also, with no policy entrepreneur or a broker stepping in to negotiate with the state and local government and complete the coupling process, that is, connect the problem with a policy and prevailing politics, the existing problem as discussed in Irving's case failed to reach the state's agenda platform. Nevertheless, as pointed out by Kingdon (1995), ideas, proposals, and issues may fade in and out but they never go away. They come back with revisions and with a new recombination or a twist. Therein lies the hope of the problem of earthquakes from fracking reaching the agenda stage in the years to come, when all the conditions have been met and policy proposals are ready to be accepted when the policy window is open.

References

Burnett, J. (2016). Texas Oklahoma divided over how to handle earthquakes linked to oil drilling. National Public Radio. Retrieved from www.npr. org/2016/11/28/503632437/texas-oklahoma-divided-over-how-to-handle-earthquakes-linked-to-oil-drilling.

City of Irving (2017). Official website. Retrieved from www.cityofirving.org/.

Dermansky, J. (2015). Texas town at center of latest earthquake swarm questions fracking impact. Desmog, blog site. Retrieved from www.desmogblog. com/2015/01/27/texas-town-center-latest-earthquake-swarm-questions-fracking-impact.

Energy in Depth. (2017a). Wastewater injection wells in North Texas rarely cause earthquakes. Retrieved from https://energyindepth.org/wp-content/uploads/2015/03/REPORT-North-Texas-Injection-and-Earthquakes.pdf.

Energy in Depth. (2017b). About EID. Retrieved from https://energyindepth.org/about/.

Frohlich, C., DeShon, H., Stump, B., Hayward, C., Hornbach, M., & Walter, J. (2016). A historical review of induced earthquakes in Texas. *Seismological Research Letters*, 87(4), 1022–1038.

Frohlich, C., Hayward, C., Stump, B., & Potter, E. (2011). The Dallas–Fort Worth earthquake sequence: October 2008 through May 2009. *Bulletin of the Seismological Society of America*, 101(1), 327–340.

Hennings, P., Savvaidis, A., Young, M., & Rathje, E. (2016). Report on House Bill 2 (2016–17) Seismic Monitoring and Research in Texas. TexNet, Bureau of Economic Geology, UT Austin.

Hornbach, M., Jones, M., Scales, M., DeShon, H., Magnani, B., Frohlich, C., Stump, B., Hayward, C., & Layton, M. (2016). Ellenburger wastewater injection and seismicity in North Texas. *Physics of the Earth and Planetary Interiors*, 261, 54–68.

Hornbach, M., DeShon, H., Ellsworth, W. Stump, B., Hayward, C., Frohlich, C., Oldham, H., Olson, J., Magnani, B, Brokaw, C., & Luetgert, J. (2015). Causal factors for seismicity near Azle, Texas. *Nature communications, 6, 1–11.*

Hornbach, M., Stump, B., & Olson, J. (2015). Presentation on Texas seismicity for the Texas House of Representatives Energy Resources Committee on May 4, 2015. Retrieved from www.smu.edu/-/media/Images/News/PDFs/Earthquake_hearing_document-04may2015.ashx?la=en.

Hornbach, M. et al. (2013). Mitigating seismicity. Presentation made on July 31, 2013. Retrieved from https://assets.documentcloud.org/documents/3220854/SMU-Presentation-June-5-Round-Table-From-SMU.pdf.

Kingdon, J. (1995). *Agendas, alternatives and public policies* (2nd ed.). New York, New York: Longman.

Kuchment, A. (2017, March 1). Are earthquakes gone from our area for good? Dallas Morning News. Retrieved from www.dallasnews.com/business/energy/2017/03/01/earthquakes-gone-area-good-scientists-try-solve-mystery.

McGarr, A., Bekins, B., Burkardt, N., Dewey, J., Earle, P., Ellsworth, W., Ge, S., Holland, A., Majer, E., Hickman, S., Rubinstein, J., & Sheehan, A. (2015). Coping with earthquakes induced by fluid injection. *Science, 347*(6224), 830–831.

Scales, M. M., DeShon, H. R., Beatrice Magnani, M., Walter, J. I., Quinones, L., Pratt, T. L., & Hornbach, M. J. (2017). A Decade of Induced Slip on the Causative Fault of the 2015 MW 4.0 Venus Earthquake, Northeast Johnson County, Texas. *Journal of Geophysical Research: Solid Earth.*

Seismological Society of America. (2014, May 1). Wastewater disposal may trigger quakes at greater distance than previously thought. *Science Daily.* Retrieved from www.sciencedaily.com/releases/2014/05/140501132628.htm.

Southern Methodist University. (2015). SMU analysis of recent earthquake sequence reveals geologic fault, epicenters in Irving and West Dallas. SMU archive. Retrieved from www.smu.edu/News/2015/earthquake-update-06feb2015.

Texas Railroad Commission. (2017a). Railroad Commission Adopts Disposal Well Rule Amendments Today. *TRRC News.* Retrieved from www.rrc.state.tx.us/all-news/102814b/.

Texas Railroad Commission. (2017b). Injection and disposal wells. Retrieved from www.rrc.state.tx.us/about-us/resource-center/faqs/oil-gas-faqs/faq-injection-and-disposal-wells/.

U.S. Census. (2015). Population and housing unit estimates. Retrieved from www.census.gov/programs-surveys/popest.html.

The Case of Transportation Problems in Fort Worth

At the beginning of the twenty-first century, when fracking first started in the Barnett Shale, no one could foresee the intensity of truck traffic that fracking would generate or the extent of damage such vehicles would exert on the existing infrastructure. As fracking gained momentum in the state in both rural and urban communities, the heavy trucks carrying materials to and from the fracking sites left behind trails of damages on highways, county, local, and farm to market roads. The evidence of damage could no longer be ignored or left unattended. It prompted several studies on transportation in areas of energy development, which have helped to reveal interesting facts and figures on truck traffic associated with fracking.

In fracking, the truck traffic varies in intensity and volume. The number of truckloads required to fracture a natural gas or an oil well tends to vary with the geology of the region, well type and depth, drilling technology, and water needs (Quiroga, Fernando, & Oh, 2012). Usually trucks and trailers used in fracking weigh from 35,000 pounds when empty to 80,000 pounds when fully loaded. According to the Texas Department of Transportation's (TxDOT) estimate, it takes 1,184 loaded trucks to stimulate production in one gas well, 353 such trucks to maintain it, and another 997 loaded trucks to refracture the wells every five years. Equating the truck traffic in terms of cars, it amounts to eight million cars and an additional two million cars to maintain a well (Barton, 2013).

The city of Fort Worth in North Texas is one of those densely populated urban communities that has experienced an increase in truck traffic as a result of a fracking boom. Here, the discovery of abundant natural gas lying underneath the city has helped to create local jobs and rendered other economic benefits. However, fracking has also brought about an increase in hazards associated with truck traffic along with safety concerns throughout the city – there has been an increase in the number of crashes between motor vehicles and truck tractors.

The City of Fort Worth

The city of Fort Worth is considered the sixteenth largest city in the nation. It is also the sixth largest city in the state of Texas and is growing fast. The city lies only 31 miles west of the metropolitan city of Dallas and is a home to 800,000 people. The historic origin of this city can be traced back to the early settlements along the Trinity River, where a camp was built followed by a fort to protect the early settlers from the native Americans. This fort was built by an American war hero named Williams Jenkins Worth, and in 1849, the United States War Department officially named this place as "Fort Worth." The name was derived from two important aspects associated with the history of the city – the presence of a "fort" and the last name of the war hero, "Worth."

In late nineteenth century, the city soon earned the nickname "Cowtown" because of its rich cattle drive history. Its strategic location in North Texas made this city a conduit to the mid-western market. With the gradual development of transportation and other means of communications in the region, the city grew very fast. After the discovery of oil and gas in the state, which predated that of energy development in the Barnett Shale, the city started attracting big oil companies. Soon they established their offices here, and their presence paved the path for the establishment of an oil stock exchange in the city.

Thereafter, the city became the favored location site for two educational institutions. A private and a state university opened their doors here and made this city a seat of learning. Public attraction sites like a museum, an art center, and a zoo added to the attractiveness and cultural value of the city. The city's close proximity to Dallas and the various amenities that it offered made Fort Worth an attractive destination for migration from various parts of the country. People from far and wide have migrated to the city and the upward trend in population growth has continued into the twenty-first century (United States History, 2017).

Fracking in Fort Worth

The city of Fort Worth is one of the first cities in the Barnett Shale that experienced an early fracking boom. At the beginning of the twenty-first century, at an opportune moment of high oil and gas prices, both local and out of state oil and gas companies adopted the advanced technology of fracking to extract natural gas from the city's vast underground reserves. Since then, fracking has spread into this urban community – fracked wells can be found even in residential areas, parks, and other places. Fracking has helped to create many local jobs in the city but from 2015 onwards, with the downturn in oil and natural gas prices, drilling has experienced a gradual slowdown here. In spite of this, it still boasts approximately 2,000 producing wells and 146 abandoned wells, located within the city limits, per the city's 2017 well count (City of Fort Worth, 2017).

During the earlier part of the twenty-first century, when local and out of state oil and gas companies started drilling for natural gas, they approached private property owners with mineral rights in the city to sign leases with them. One of the companies, Chesapeake Energy based in Oklahoma, made the biggest splash in the city with its monetary contributions and sponsorships. It contributed $100,000 to build a memorial for policemen and firefighters and donated another $1 million for a new science and history museum in the city. It also spent $1 million to start the Barnett Shale Endowment Fund. The company also went as far as sponsoring the city's most attractive holiday event, which was renamed the Chesapeake Energy Parade of Lights (Elkind, 2014).

In pursuit of mineral rights held by city residents, a rivalry ensued between a local oil company called XTO and Chesapeake Energy. Both the companies tried to outbid each other in their offer of initial payments to private property owners for signing of leases. This created a bidding war among mineral rights owners to sign a lease with the company that offered them the maximum money along with a promise of royalty payments. Even the city itself could not resist the lure of money. It leased public land (parks and cemetery sites) for natural gas drilling, and so did the local newspaper company, under its printing plant. With the surge in signed leases, the pace of drilling for natural gas gained momentum within the city limits. Drilling rigs appeared in various parts of the city including residential neighborhoods, country clubs, and cemeteries. Along with the rigs, there was a concurrent increase in truck traffic on local and farm to market roads. Heavy trucks transporting materials to and from the fracking sites caused damages to the city's existing infrastructure, which was not built to handle such heavy loads but rather meant for pickup trucks and farm tractors.

Increase in Truck Traffic

The city of Fort Worth just like any other fracking site in the state experienced an increase in three types of truck traffic: trucks associated with seismic activities, construction, and production. The seismic trucks were used in seismic explorations to generate vibrations or seismic waves that were transmitted underground. The seismic waves were reflected back to the surface and were then analyzed by geologists to locate the "sweet spots" or potential large deposits of natural gas underneath the surface layer. Since the seismic trucks are usually very heavy, weighing 40,000 to 65,000 pounds, their use is often subject to certain restrictions in urban areas as they are known to cause untold damages to old underground pipes.

At the fracking sites, the five stages of the well development process – site preparation, rigging up, rigging down, drilling, and hydraulic fracturing – generated construction truck traffic. The construction trucks used the federal and state highways, farm to market roads, local and county roads and spurs, beltways, loops, and business roads. The production trucks were mostly involved in the transportation of salt or wastewater to the injection sites.

In Fort Worth, the truck traffic generated by fracking undoubtedly added to the existing fleet of truck traffic in the city. This is evident from Table 6.2 and the maps shown below. Though it is difficult to pinpoint the exact increase in the number of trucks that are associated with fracking, the overall increase in the city's truck traffic can be seen from TxDOT's truck traffic maps of 2005 and 2014. Much of this increase in truck traffic can be attributed to fracking. Furthermore, the increase in truck traffic has coincided with an increase in the number of crashes involving motor vehicles and truck tractors in the city (see Figure 6.9).

Table 6.3 Truck Traffic Produced by One Single Well in Barnett Shale

Type of Vehicle	Number of Axles	Loaded weight in pounds	One-way trips (per well site)
Work over Rig	5	80,000	2
Work over Rig	5	80,000	2
Tank Truck	5	80,000	70
Water Tanker	5	80,000	685
Water Tanker	5	80,000	214
Bob Tail	5	80,000	24
Production Tank Truck	5	80,000	353

Source: Adapted from Table 2 of Impacts of Energy Development on Texas Roads by Li and Mikhail, a paper submitted for publication and presentation at the 9th International Conference on Managing Pavement Assets in Washington, D.C. August 2014.

Table 6.4 Daily and Truck Traffic in Fort Worth District

All Similar Roadways in the Fort Worth District

Year	Average Daily Traffic	Average Truck Percent	Average Truck Traffic
2000	954	14	134
2001	949	14	133
2002	1,034	15	155
2003	1,055	14	148
2004	1,072	11	118
2005	1,147	13	149
2006	1,214	13	158
2007	1,258	19	239
2008	1,091	18	196
Total	1,086	15	159

Note
Fort Worth District includes the City of Fort Worth.

Recent studies on truck transportation in areas of energy development in the state have revealed that drivers tend to show a preference for those routes that enable them to avoid weigh stations, law enforcement personnel, and hazardous road conditions (Quiroga, Fernando, & Oh, 2013). This leads to greater use of local and farm to market roads, and such a tendency is not uncommon in Fort Worth. Even though there are designated truck routes connecting the well sites, drivers also use local and farm to market roads. The unprecedented increase in the number of trucks on these roads has caused damages and reduced the life expectancy of these roads.

As per the city's gas drilling ordinance, injection wells cannot be located within the city limits. As a result, the saltwater generated as a backflow during

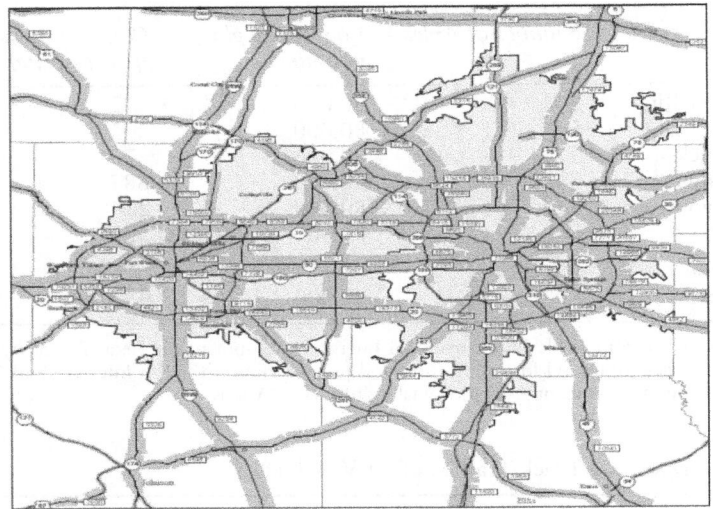

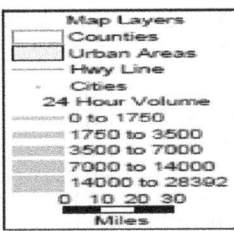

Figure 6.10A Truck Traffic in Dallas – Fort Worth Area in 2005.

Source: 2005 Texas Truck Flow Band Map prepared by Texas Department of Transportation, Transportation Planning and Programming Division in cooperation with U.S. Department of Transportation.

the fracking process has to be hauled outside the city for storage in injection wells that are subject to regulations by the Texas Railroad Commission (Prozzi Grebenschikov, Banerjee, & Prozzi, 2011). Since injection wells have a limit of 25,000 gallons of wastewater, it takes 243 truckloads to transport wastewater from a fracking site to a disposal well. Further, the truck traffic's intensity is not the same throughout the day. A period of heavy truck activity is often followed by a period of lull, which makes it difficult to capture reliable truck traffic data in any urban community where there is fracking.

In the handling of wastewater, a difference exists between that of Fort Worth located in Barnett Shale and a city in the Permian Basin. In the Permian Basin, which is noted for its oil production, nine barrels of saltwater are produced per barrel of oil. The produced saltwater is reinjected

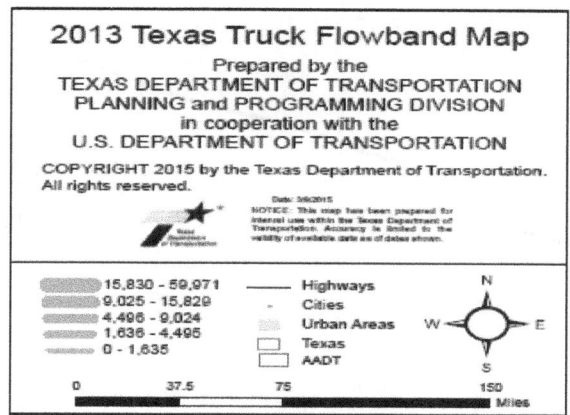

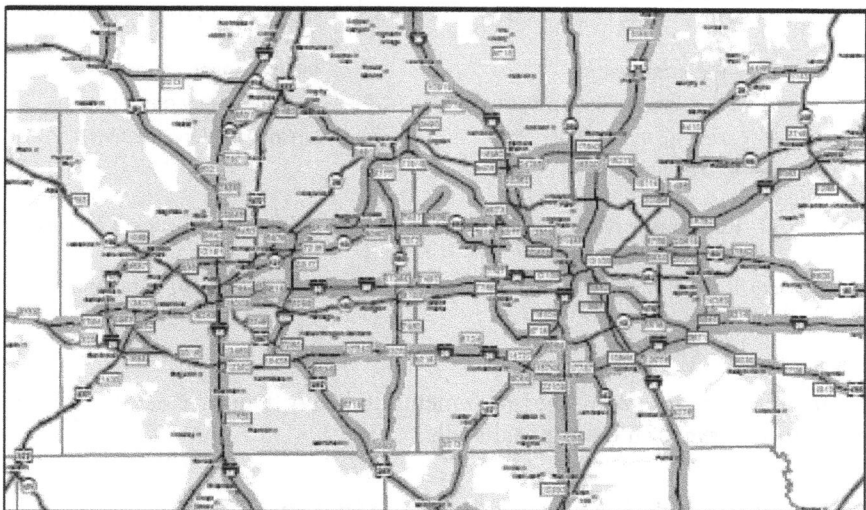

Figure 6.10B Truck Traffic in Dallas – Fort Worth Area in 2013.

Source: 2013 Texas Truck Flow Band Map prepared by Texas Department of Transportation, Transportation Planning and Programming Division in cooperation with U.S. Department of Transportation.

into the drilling well to increase pressure and stimulate the flow of oil (Prozzi et al., 2011). The reuse of saltwater helps to minimize the distance traveled by trucks to dispose of saltwater into injection wells. This translates to less wear and tear on roads and money savings from costly repairs and maintenance costs of local roads. Such reuse is not possible in the Barnett Shale, and as a result, the local roads cannot be spared from

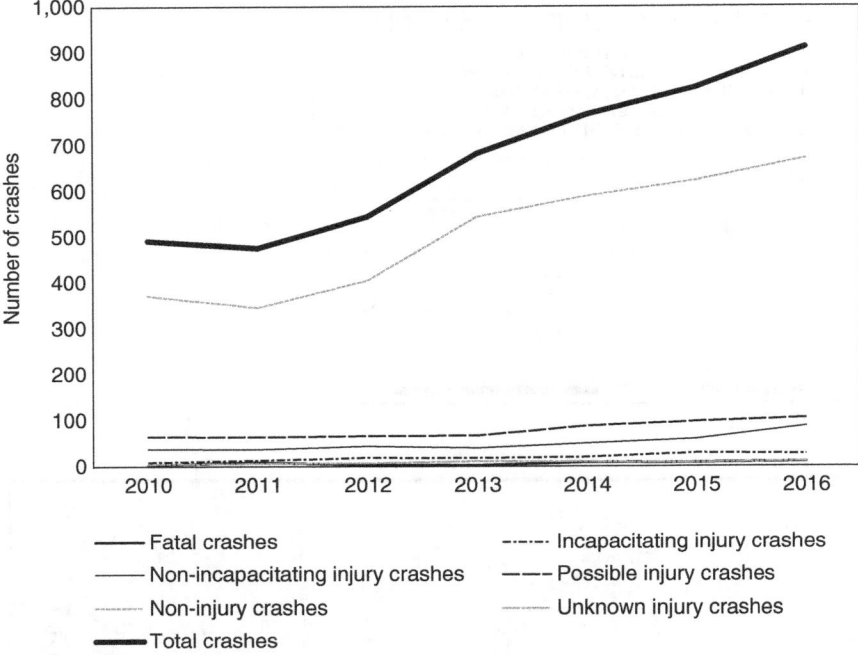

Figure 6.11 Reportable Motor Vehicle Traffic Crashes Involving Truck Tractors in Fort Worth, Texas 2010–2016.

Source: Data for the graph was obtained from the Texas Department of Transportation, 2010–2016.

imminent damages from the movement of heavy trucks carrying wastewater to the injection or disposal wells.

The Community's Response to Increase in Truck Traffic

The increase in truck traffic in Fort Worth, along with other negative externalities associated with fracking, have made city residents develop a negative attitude toward it, evident from a survey (Anderson & Theodori, 2009). Even though the city's gas ordinance requires an oil and gas company to submit a plan for designated routes to access the fracking site, in recent years the city has approved residential developments near the gas wells. This has led to the mixing of residential traffic with truck traffic at certain intersections where the designated truck route either skirts or runs through residential developments for small distances. The appearance of truck traffic closer to homes has made the local residents leaving or entering their neighborhoods exercise extra caution in the handling of truck traffic on local roads. The truck traffic poses a serious safety concern and has prompted the city to call for a review of designated truck routes since these routes were approved a decade ago at a time when no such residential developments existed.

The various types of road damages generated by trucks have prompted local residents to file numerous complaints with the city for road repairs. The city has a cost sharing agreement with oil and gas companies operating within the city limits for the repair of designated truck routes. Unlike some cities in the Barnett Shale, it does not require operators to pay a road impact fee at the time of receiving a permit for the drilling of gas wells within city limits. Instead, operators are required to submit a video of the designated truck route along with a road repair agreement. This agreement enables the city to recover much of the repair costs associated with damages caused by heavy trucks that are entering and leaving the city.

In assessment of an operator's share of repair costs for damages, visual comparison of the road condition at the time of permit approval and at a later period proves helpful. The city's traffic engineers usually assess the extent of damage and negotiate a deal with operators for cost sharing in road repairs. For example, an operator's share of repair costs for truck related damages on an old road is likely to be low. In such an assessment, the age of the road and normal wear and tear are factored and reduce the operator's cost burden. Whereas the opposite is true in assessment of truck related road damages on newly built roads.

Some operators try to keep their share of repair costs low by taking pictures of truck routes periodically and submitting them to the city for record keeping. Especially when designated truck routes are also accessed by construction trucks for transportation of materials to residential development sites, the archived pictures come in handy. They help the city to make a fair decision in assessment of the operator's share of repair costs for road damages rendered by trucks. Over the last few years, the slump in oil and gas prices and concurrent slowdown in fracking activities have led to the collapse of repair cost sharing agreements in some cities of the state (Li & Mikhail, 2014). This is not the case with Fort Worth. The city's cost sharing agreements made with the operators still remain valid today.

With reference to city residents' resistance to truck traffic, it has manifested in the form of complaints only. Aside from the lodging of complaints for road repairs and sightings of trucks on residential streets day and night, city residents have not taken any other actions to curb truck traffic within the city. Neither does there exist a citizen group to protest at the movement of truck traffic within the city nor has any initiative been undertaken by citizens to demand changes in local ordinance or policy to exercise more control on the movement of truck traffic within the city.

Understanding the Community's Weak Opposition to Fracking

In Fort Worth, the community's opposition to fracking has been relatively weak in comparison with that of Denton and Irving. Though people complained about traffic related problems within the city, no large scale citizen driven initiatives were undertaken to attract attention and gain prominence to the problem at the state level. One plausible explanation might be the city's timely intervention. By entering into agreements with the oil and gas operators, the city was able to partially control the movement of heavy trucks and extract some compensations for the damages incurred to its infrastructure.

Another plausible factor that could have undermined the scale and strength of opposition is the prevailing individualist culture of the city and state. It made individuals more willing to accept the risks associated with the increase in truck traffic to and from the oil and gas drilling sites. The royalty payments that community residents received from leasing their oil and gas rich properties and the benefits accrued from the investments made in the city by the oil and gas companies made the community members realize that they were in the domain of economic and societal gains since any deviation from it would cause economic hardships and associated losses, which most individuals are reluctant to accept; this might be a reason for the city residents' weak opposition to the negative externalities from fracking related activities in the city.

References

Anderson, B. & Theodori, G. (2009). Local leaders' perceptions of energy development in the Barnettt Shale. *Southern Rural Sociology, 24*(1), 113–129.

Barton, J. (2013). Presentation to House Appropriations Subcommittee Committee on Budget Transparency and Reform. Texas Department of Transportation. Retrieved from https://ftp.dot.state.tx.us/pub/txdot-info/energy/presentation_031113.pdf.

City of Fort Worth. (2017). Gas well drilling. Retrieved from http://fortworthtexas. gov/gaswells/.

Elkind, P. (2014). The fracking hangover in the heart of oil and gas country. Retrieved from http://fortune.com/2014/05/31/fracking-hangover/.

Li, Z. & Mikhail, M. (2014). The Impacts of Energy Development on Texas Roads, a paper submitted for publication and presentation at the 9th International Conference on Managing Pavement Assets in Washington, D.C. August 2014.

Prozzi, J., Grebenschikov, S., Banerjee, A., & Prozzi, J. (2011). Impacts of energy developments on the Texas transportation system infrastructure. Technical Report, number 0-6513-1A. Center for Transportation Research, University of Texas at Austin.

Quiroga, C., Fernando, E., & Oh, J. (2013). Energy developments and the transportation infrastructure in Texas: Impacts and strategies. A technical report from the Texas Transportation Institute.

United States History. (2017). History of Fort Worth, Texas. Retrieved from www.u-s-history.com/pages/h3888.html.

The Case of Water Shortage in Carrizo Springs

In the fracking of shale rocks for oil and natural gas production, millions of gallons of water are used. The water's source can be either surface, underground, recycled, municipal, or a combination of two or more types. From the rampant fracking of shale formations in the state, two things have become evident: water consumption varies from one region to another and oil production requires less water than natural gas. The amount of water required actually depends upon the geology of the basin, well spacing, fracturing stages, technology used, whether stimulation is required, and the quantity of flowback water that is generated that can be reused in fracking.

In the Barnett Shale, mostly surface water is used in fracking and at times of drought, the TCEQ can suspend the use of surface water for fracking (Kurth, Mazzone, & Mendoza, 2012), while in the arid parts of the state, both ground and brackish types of water are used. The groundwater obtained from the wells is less regulated than the surface water. Oil and gas companies often buy or lease lands to drill water wells to meet their requirements. It can cost them $1 million to drill a deep well or $70,000 to $80,000 for a shallow well (Galbraith, 2013a).

In the use of groundwater for fracking, the oil and gas companies do not need to obtain permits from the following agencies: the Texas RRC, most Groundwater Conservation (GCD) districts, river authorities, special districts, priority groundwater management areas, water utilities, and counties of the state. The only time a permit is required from a regulatory agency is when surface water use is involved in fracking, where a permit has to be obtained from the TCEQ. The management of groundwater in Texas is typically done by the GCDs. There are nearly 100 GCDs in Texas who enforce rules for the distribution and use of water based on fairness, property rights, public interest, and the district's management plan. If an oil and gas company can show that the groundwater is to be solely used for oil and gas exploration and the operator will be drilling the water well, then no permit is required from GCDs under the state's water code Section 36.117. Even though permits are not required by oil and gas companies under specific circumstances, an exempted well has to be registered with the GCD and precautions must be taken to prevent contamination of groundwater. If groundwater from an exempted well is transported outside the water district for use in fracking elsewhere, the GCDs require the oil and gas company to pay for the production and export fees (Kulander, 2013).

In most parts of the state, there are no restrictions on the withdrawal of groundwater. The groundwater used by oil and gas companies for fracking is neither monitored nor requires any reporting to the state. This makes it difficult to obtain accurate data on water consumption for fracking in the state. Only a few operators maintain a log on their water use for fracking. They are sometimes reluctant to share that data because of the sensitive nature of the information in it (Nicot et al., 2011).

According to Ceres, a Boston-based non-profit sustainability organization, the highest water consumption in fracking is in Texas. This news comes as no surprise. The state leads the nation in the number of fractured wells in its massive stretches of shale formations in various parts of the state. The use of groundwater in fracking poses some serious concerns. The withdrawal of large amounts of groundwater poses the risk of resource depletion over time since it takes a longer time to replenish an aquifer. Other risks include land subsidence, reduction in surface water flows, scarcity of water, and challenges in distribution of water in those local communities that heavily rely on groundwater (Freyman, 2014).

The flowback water that gushes back in the fracking process is a mixture of ground and injected water. It is often recycled for reuse in fracking in the state, but this recycled water constitutes only a small fraction of water used in fracking. Recycled water is usually expensive but in the arid areas of the state there exist no other options but to use it. For example, in the Anadarko Basin, located in the Texas panhandle, 20 percent of the water used in fracking is recycled. In the Barnett Shale and Permian Basin, only 5 percent of the water used in fracking is recycled while little or no recycled water is used in the Eagle Ford Shale (Nicot & Scanlon, 2012). In the latter case, the geology of the region is not conducive to the production of flowback water.

In the Eagle Ford Shale, approximately 90 percent of the water used in fracking is obtained from underground wells while the remainder is brackish water (Nicot, 2013; Freyman, 2014). In North Texas, municipal water is also used in fracking but it constitutes a relatively small amount. The city of Arlington acts as the supplier of municipal water to the local oil and gas operators in the Barnett Shale (Nicot et al., 2011).

The demand for water also depends upon the intensity of fracking. When there is an upward trend in the intensity of fracking, it leads to higher water consumption while a downward trend has the opposite effect. This intensity of fracking in turn is usually determined by the market price for oil and natural gas. With the rapid decline in oil and gas prices in 2016 there was a consequent slowdown in the intensity of fracking in the state. From 2017 onwards, with the gradual increase in the oil and gas price after hitting the bottom the previous year, the intensity of fracking has gained momentum in the state. Also, the focus has shifted from Barnett Shale, the initial production site, to the Permian Basin, which is now producing the largest amount of oil and natural gas in the state.

Allegations of Heavy Water Use

Many environmental organizations in the state have blamed the oil and gas companies for their heavy water use in fracking. The industry has retaliated by pointing out that their water use is only a small fraction of total water consumption in the state. It is less than 1 percent of total water used in the state, even less than the total amount required in the watering of lawns in the urban areas (Galbraith, 2013a).

A fact check on the industry's water use for fracking has shown that from 2001 to 2010, the mining industry, including oil and gas production, did use the lowest amount of water in the state in comparison with that of irrigation, municipalities, manufacturing, power generation, and livestock farming. But in the category of mining by itself, fracking constituted the second highest amount of water use (approximately 23 percent) after the crushed stone industry (Nicot & Scanlon, 2012).

The oil and gas industry's framing of the issue of water consumption in fracking at the state level has only helped to distract from the seriousness of this issue and quell public fear. Since fracking is not conducted at all drilling sites and only in the shale formations of the state, the problem is more local in nature than statewide. Water used in fracking is more of a concern among rural and urban communities that are located on top of the shale deposits and in arid parts of the state. In the Barnett Shale, fracking accounted for 10 percent of water use in Tarrant County and 19 and 20 percent in the Wise and Johnson counties of North Texas in 2013 (Freyman & Salmon, 2013).

In other water-stressed parts of the state and at times of drought, the withdrawal of large amounts of water for fracking has meant less water is available for local irrigation and ranching. It poses a serious concern to local farmers and ranchers in the state, whose livelihoods depend upon the availability of local water supply. With the increase in the number of

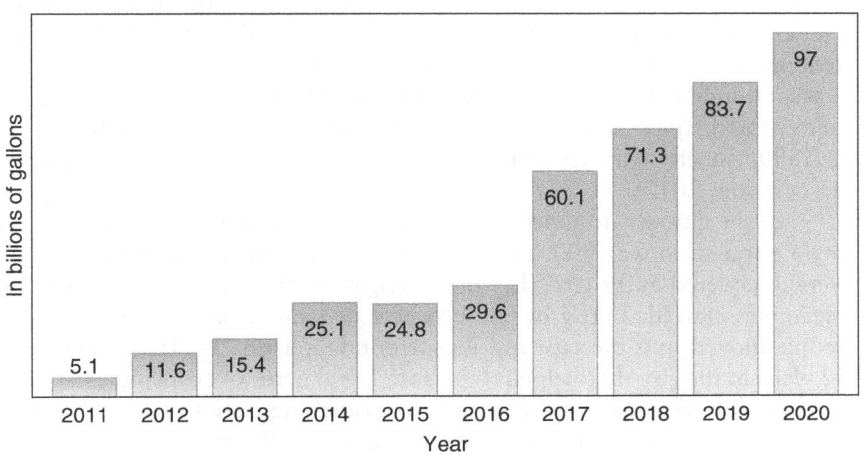

Figure 6.12 Water Use in Fracking in Permian Basin, Texas.

Source: IHS Markit and Houston Chronicle.

Note
Amounts for 2019 and 2020 are estimates.

exploration and production sites of oil and gas in the Permian Basin, water consumption in fracking has only gone up over the years.

The fear of water scarcity has continued to persist since fracking began in the state. To ease fracking's demand for water in the arid Permian Basin, a land baron and oil man in West Texas has made public in 2017 a plan to pump 5.4 million gallons of water a day from an aquifer lying beneath his 140,000 acres ranch and transport it 60 miles to a fracking site. The plan has alarmed local residents, farmers, ranchers, and environmentalists. They fear the depletion of water will affect them adversely and even dry up a popular spring in the Balmorhea State Park. So, locals not only oppose the plan but are also ready to launch a lawsuit if it is finalized (Hunn, 2017).

In the Eagle Ford Shale, where the city of Carrizo Springs is located, nearly a quarter of the water available in this arid part of the state is used in fracking. The Winter Garden Water Conservation District, which serves the counties of Dimmit where Carrizo Springs is located, La Salle, and Zavala, has no restrictions on withdrawal of groundwater for fracking. As a result, during a period of drought that struck the region between 2012–2013, the local farmers and ranchers experienced a drop in the water level of their wells. This resulted from the heavy pumping of groundwater by local oil and gas companies closer to the existing wells for use in fracking in the Eagle Ford Shale.

History of Carrizo Springs

The city of Carrizo Springs lies southwest of San Antonio. It is less than 50 miles north from the Mexican border and located in the Dimmit County. The city derived its name from the many springs that were found in the area and the Spanish name of the cane grass that grew around the springs. It was founded in 1865 by settlers from the Atascosa County, and the city enjoys the distinction of being the oldest and the largest city in the county. In 1880, the city was designated as the county seat (Texas State Historical Association, 2017).

Since the city was founded, its population has grown slowly. In 1885, the city's population was 900. In 1900, when wells with clean and pure water were discovered in the city, they made irrigation possible, and the farming of vegetables started. Lured by the prospect of gains from agriculture, many people moved into the city and its surrounding areas. By 1904, there were 30 wells in the city that helped to irrigate 1,000 acres of cropland.

With the expansion of farming and ranching activities, more people moved into the city and its population reached 1,200 in 1916. In 1928, the city experienced another surge and its population reached 2,500 people. In 1984, the county's only newspaper, *Carrizo Springs Javelin*, opened its printing office in the city, as did a local radio station. The gradual increase in business along with farming and ranching activities enhanced the attractiveness of this city. More people moved into Carrizo Springs and the city's

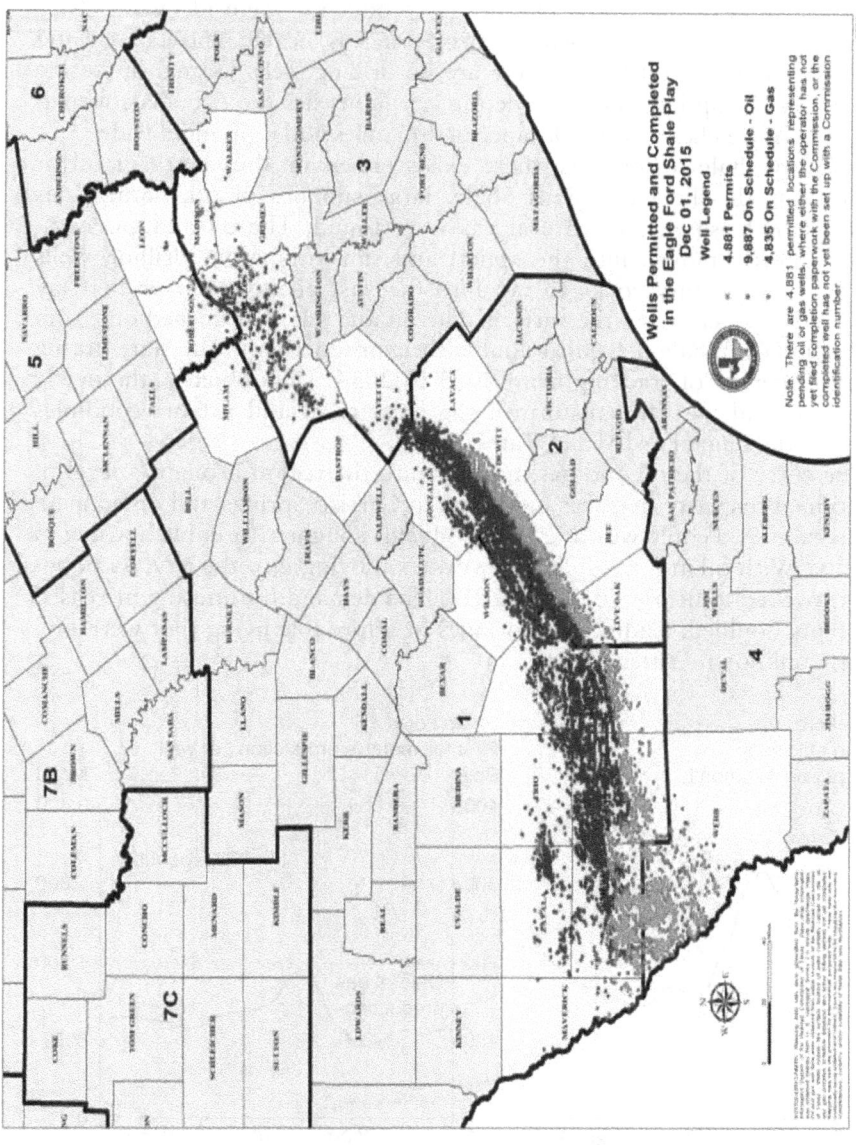

Figure 6.13 Distribution of Oil and Gas Wells in the Eagle Ford Shale of South Texas.

Source: Energy Information Administration (EIA), USA.

population reached 5,655 people in 2000 (Texas State Historical Association, 2017).

In 2008, the discovery of oil and gas in the Eagle Ford Shale simply helped to further enhance the importance of this small city. In the Eagle Ford Shale, two core areas of maximum oil and gas production were identified. They included Dimmit County, where Carrizo Springs is located, and La Salle and Zavala counties. Even though there are no drilling wells located inside the city of Carrizo Springs, just outside the city limits lies the rich shale formation, which stretches across a distance of 400 miles and is 50 miles wide.

In many shale formations, there exists either an abundance of oil or natural gas. In the Eagle Ford Shale, large supplies of oil, natural gas liquids (condensate), and natural gas were found. The oil and gas companies quickly moved into the region and sunk numerous drilling wells into the sedimentary rocks to tap into the vast energy reserves that lay thousands of feet below the surface. The oil and gas that gushed out from the Eagle Ford Shale helped to double their production in the state during the peak period of fracking from 2013 to 2014. Undoubtedly, the newly found mineral wealth transformed the local rural and urban economies found in the Eagle Ford Shale (Yates, 2014).

The entry of the oil and gas industry into the region brought a level of economic prosperity that was unknown to Carrizo Springs and other small municipalities. People working in the oilfields sought affordable housing in the city. With oil and gas industry workers moving into the city, its population swelled to nearly 40,000 in 2013. The demand for housing increased and rents doubled, leading to shortages in affordable living that were previously unknown (Yates, 2014).

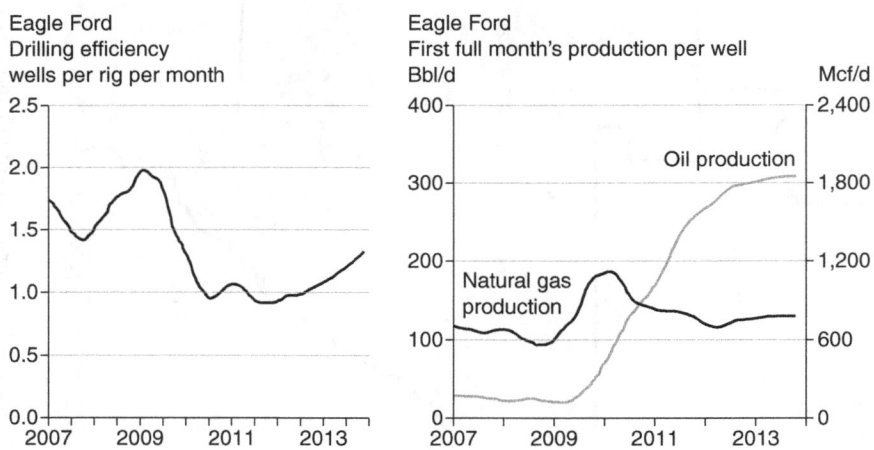

Figure 6.14 Drilling Efficiency and Production of Oil and Natural Gas in Eagle Ford Shale.

Source: Energy Information Administration (EIA), USA.

Table 6.5 Economic Changes in Eagle Ford Shale in the Dimmit County since 2007 when Fracking Started

	Annual Percentage Change In Employment			*Annual Percentage Change in Income*		
County	2001–2006	2006–2010	2010–2013	2001–2006	2006–2010	2010–2013
Dimmit	−0.02	3.44	22.93	5.26	7.79	18.30

Source: Table 5.1 and 5.2 of the Eagle Ford Shale Report from the Economic Impact of Eagle Ford Shale, Institute for Economic Development, University of Texas, San Antonio.

Table 6.6 Use of Surface and Groundwater in Dimmit County

Water Use	Source	2013	2012	2011	2010	2009	2008	2007	2006
Municipal	GW	2,447	2,432	2,400	2,175	2,302	2,267	1,805	2,382
	SW	0	0	0	0	0	0	0	0
Mining	GW	9168	7547	3453	1554	824	94	0	0
	SW	1018	640	255	115	7	0	0
Irrigation	GW	4,433	5,894	5,570	7,170	7,831	6,191	3,041	4,507
	SW	87	203	52	45	226	878	363	1,500
Livestock	GW	155	203	232	228	233	258	217	294
	SW	155	203	233	228	234	259	216	294

Source: Data obtained from Estimated Historical Water Use and 2012 State Water Plan Dataset: Winter Garden Groundwater Conservation District.

Note
Groundwater (GW); Surface water (SW) measured in acre fee/year.

Since fracking started, the fracked wells have produced 40,276 billion barrels of oil and 214,126 Mcf (thousand cubic feet) of natural gas (Texas Drilling.com, 2017). Fracking brought about an increase in Dimmit County's revenue. The unemployment rate dropped from 12 to 4 percent (Hennessey-Fiske, 2014) when the total number of energy related jobs almost doubled from 603 to 1187 in 2010. In 2013, as fracking reached its peak, local employment numbers in oil and gas companies increased to 1,533 (Center for Community & Business Research, 2014).

In 2016, the economy of Carrizo Springs went from boom to bust when natural gas and oil production slowed down with the global decline in prices. As the global market recovered, production continued in the Eagle Ford Shale. In 2017, there were 59 producing leases, 16 producing operators, and 1,186 wells near Carrizo Springs.

Water Shortages in Carrizo Springs

One of the downsides of economic prosperity in Carrizo Springs was the water shortage. With millions of gallons of groundwater used for oil and gas production in the Eagle Ford Shale, it made Dimmit County's aquifer level drop by 100 to 300 feet (Freyman & Salmon, 2013). According to a Ceres water report (Freyman, 2014), Dimmit County alone consumed over four billion gallons of water in fracking, the highest amount in the state and the nation. The nearby counties of DeWitt and LaSalle also experienced similar water shortages and their aquifer levels dropped by 100 to 300 feet.

The high intensity of fracking during the period of 2012–2013 that coincided with a period of drought in the region posed serious concerns in local water resource management in the Eagle Ford Shale. The Wintergarden Groundwater Conservation District (WGCD), serving Carrizo Springs and other small municipalities in the Eagle Ford Shale, allowed the oil and gas companies to withdraw as much groundwater as they needed for fracking without a permit. This is not uncommon in the state. The Railroad Commission allows unlimited use of groundwater for oil and gas production whereas the use of surface water for such purposes requires permission from the TCEQ (Rahm, 2011).

The WGCD made exceptions to the permit requirement for wells under Section 9 and Rule 9.1 of its management plan (2007), which included the following categories of wells under its jurisdiction.

1   A well used to supply water necessary for mining as authorized by the Railroad Commission of Texas under Chapter 134, Natural Resource Code.
2   A well used to supply water solely for a drilling rig that is actively engaged in drilling or exploration operations as authorized by the Railroad Commission of Texas if;

   a   the person holding the permit issued by the Railroad Commission of Texas is responsible for the water well; and
   b   the water is located:

      i   on the lease on which the drilling rig is located
      ii   within the boundaries of the field in which the drilling rig is located or
      iii in close proximity to the rig.

In the water district's 2016 management plan, the exemptions have remained intact. All exempt wells in the district have to be registered for identification purposes, and registration also requires a payment of a fee to WGCD. Some of the information that is required in the registration process includes the proposed location of well, conditions for exemption, depth of well, well completion record, and other relevant information requested by the board members (Wintergarden Groundwater Conservation District, 2007).

To meet the high demands for water in fracking in the Eagle Ford Shale, the oil and gas companies also bought water from the local farmers and

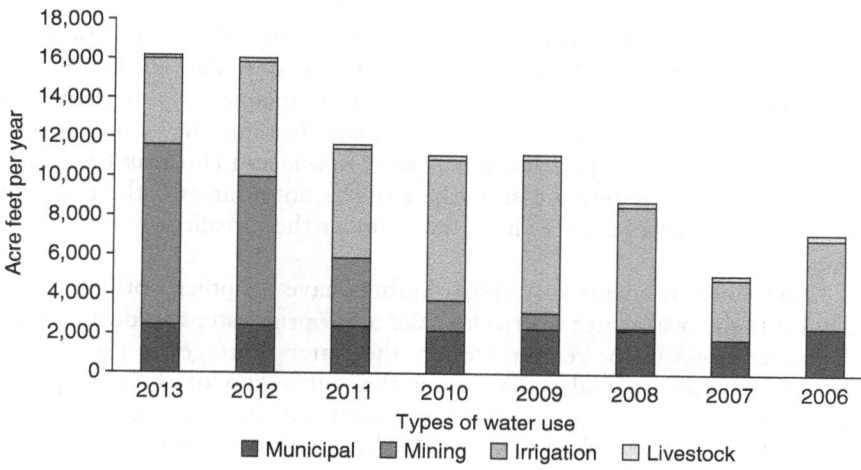

Figure 6.15 Groundwater Consumption in Dimmit County.

Source: Graph based on data obtained from Texas Water Development Board's, "Estimated Historical Water Use" and 2012 State Water Plan Dataset: Winter Garden Groundwater Conservation District.

ranchers, who sold water from their private wells at the price of 75 cents per barrel (each containing 55 gallons of water) to supplement their income.

The high consumption of water in fracking is evident from the statistics on local water use in this region. State data on groundwater use in Dimmit County has shown a gradual increase in the category of mining from 2007 onwards. Prior to 2007, there existed no groundwater consumption in that category. From 2008 onwards, groundwater use in mining has shown a gradual increase while the water available for livestock and irrigation has shown a consequent decline during that same time period.

The drought that started in 2011 in the Eagle Ford Shale region and lasted until 2013–2014 affected Carrizo Springs. The city's 21 inches of annual rainfall seemed insufficient to replenish the nearby aquifers and wells. Irrigation and ranching suffered a setback in the city (Galbraith, 2013b).

In 2012, when the price of natural gas dropped from $12 to $2 per thousand cubic feet, its production stopped. Oil and gas companies shifted their attention to oil production (Tunstall, 2014). In 2015, when the glut in the global market led to a further downward spiral in the price of a barrel of oil, the region received another heavy blow to its oil production. Thousands of oil- and gas-related jobs were lost and the consequences could be felt as the local economy went from boom to bust.

Community's Responses to Water Shortage

The water stress in Carrizo Springs has been mainly felt by local ranchers and to a smaller extent by farmers, according to state data on local water use. Surprisingly, only a few people have complained about the water shortage to the WGCD, even though it made headline news in the local and some state newspapers like the *Texas Tribune* and *Houston Chronicle*. In the absence of public pressure, the city has not addressed the problem, since all water related issues have fallen under the jurisdiction of the local water district.

Community residents of Carrizo Springs have no other options but to rely on their local water district to take appropriate steps in dealing with the water stress in the region. Despite the water shortages in the district, WGCD's failure to make changes in the status quo of the exemptions granted to the oil and gas industry can be partly attributed to the ambiguities in the state's water code. Since the code's adoption preceded the practice of fracking, confusion continues among water districts about whether to treat fracking as part of drilling and exploration activities or as a production process (Galbraith, 2013b). Unless a clarification is made by the state, many water districts like WGCD will not make changes to their policies.

Alternatively, WGCD has sought other avenues to manage and conserve water in its jurisdiction. It has opposed the siting of injection or saltwater disposal wells in the district, introduced educational programs aimed at water conservation, and encouraged school students to collect rainwater for later reuse. Unlike in other districts, WGCD's water management plan has not called for the recycling of water used in fracking since the geology of the region hinders the production of flowback water.

To protect and conserve water, the district has developed several mutually beneficial partnerships. It has entered into a partnership with Texas A & M to increase efficiency in irrigation practices and to study rainwater absorption in brush land. Partnerships also exist with the Nueces River Authority and various groundwater districts to educate people on prevention of pollution and conservation of water. Further, to optimize and conserve the groundwater of Carrizo aquifer, the district collaborates with the Southwest Research Institute to study groundwater use in the district and to fingerprint the Carrizo aquifer to identify the sources of pollutants, if any.

Understanding the Community's Mild Opposition to Water Shortages

The opposition to water shortages in the City of Carizzo Springs is much milder when compared to that of Denton, Irving, and even Fort Worth. In accounting for its mild nature, two factors are likely to have played an important role. They include the increase in the number of jobs that were made available by the oil and gas companies in the region and the effects of the water crisis being limited to a small section of the city's population, mainly the farmers and cattle ranchers. Since other stakeholders in the city experienced no disruption to their municipal water supply, which showed little or no change, most city residents did not participate in the opposition. Instead, as direct beneficiaries of job growth in the region and the boom in the local economy, they welcomed the entry of oil and gas companies in this rural part of the state. Coupled with the prevailing individualist culture of the city and the state, the city residents developed a favorable attitude toward oil and natural gas production. Perhaps it made them overlook the negative externality of water shortage associated with fracking that took place outside the city limits.

The members of the agricultural community who experienced a decline in their groundwater supply due to the withdrawal of huge amounts of water for fracking from nearby underground wells voiced opposition to the local water shortage. Specially at a period of drought, the permits given by the water conservation district to the oil and gas companies to dig underground wells and withdraw unlimited amounts of groundwater amounted to some diversion of water from traditional irrigation practices to the newly added economic activity of oil and gas production. This posed a risk to the livelihood of the farming community. It made them protest against the heavy use of water in fracking for oil and natural gas production, but their opposition failed to garner much public attention or prominence both at the water district and state levels. Instead, it created awareness among watchdog groups in and out of the state and those concerned state residents who realized that the time had arrived for changes in the state's water code. Its ambiguous nature would only lead to the rapid exhaustion of this scare resource in a state with a growing population and increasing demands for water.

The other factors that undermined the strength and scale of opposition to water shortages in Carizzo Springs were the lack of empowerment in the rural community, and little or no intervention and support from environmental groups within and outside the state. With no strong advocates for changes in the state's water code and a requirement of recycling of water used in fracking, the prospect of such policy changes is distant.

References

Center for Community & Business Research. (2014). Economic impact of the Eagle Ford Shale. A Report from the Institute for Economic Development of University of Texas, San Antonio. Retrieved from http://iedtexas.org/wp-content/uploads/2014/09/2014_EFS_Release_Oct.pdf.

Energy Information Administration. (2013). Drilling efficiency is a key driver of oil and natural gas production. Retrieved from www.eia.gov/todayinenergy/detail.php?id=13651.

Freyman, M. (2014). Hydraulic fracturing and water stress: Water demand by the numbers. A Ceres Report.

Freyman, M. & Salmon, R. (2013). Hydraulic fracturing and water stress: Growing competitive pressures for water. A Ceres Report.

Galbraith, K. (2013a, March 13). Ambiguities Reign in Regulations for Groundwater Fracking. *Texas Tribune*. Retrieved from www.texastribune. org/2013/03/13/fracking-groundwater-rules-reflect-legal-ambiguiti/.

Galbraith, K. (2013b, March 8). Fracking increase spurs fears over water use. *Texas Tribune*. Retrieved from www.texastribune.org/2013/03/08/texas-water-use-fracking-stirs-concerns/.

Hennessey-Fiske, H. (2014). Fracking brings oil boom to south Texas town, for a price. *Los Angeles Times* online, February 15, 2014. Retrieved from http://articles.latimes.com/2014/feb/15/nation/la-na-texas-oil-boom-20140216.

Hunn, D. (2017, August 2). As the oil patch demands more water, West Texas fights over a scarce resource. *Houston Chronicle*, A1 & A15. Retrieved from www.houstonchronicle.com/business/article/As-the-oil-patch-demands-more-water-West-Texas-11724100.php.

Kulander, C. S. (2013). Shale oil and gas state regulatory issues and trends. *Case Western Reserve Law Review*, 63, 1101–1140.

Kurth, T. E., Mazzone, M. J., & Mendoza, M. S. (2012). American Law and Jurisprudence on Fracing—2012. *Rocky Mountain Mineral Law Foundation Journal*, 58(4).

Nicot, J. (2013). Hydraulic fracturing and water resources: A Texas study. Search and Discovery Article # Adapted from oral presentation given at AAPG Geoscience Technology Workshop, Solving Water Issues in the Oil Field: Using Geology and More, Fort Worth, Texas, February 26–27, 2013.

Nicot, J. & Scanlon, B. (2012). Water use for shale gas production in Texas, US. *Environmental Science & Technology*, 46(6), 3580–3586.

Nicot, J., Reedy, R., Costley, R., & Huang, Y. (2012). Oil and gas water use in Texas: Update to the 2011 mining water use report. Prepared for Texas Oil and Gas Association by Bureau of Economic Geology, The University of Texas at Austin.

Nicot, J., Hebel, A., Ritter, S., Walden, S., Baier, R., Galusky, P., Kyle, R., Symanka, L., & Breton, C. (2011). Current and projected water use in the Texas mining and oil and gas industry. A report prepared for Texas Water Development Board by Bureau of Economic Geology, The University of Texas at Austin.

Rahm, D. (2011). Regulating hydraulic fracturing in shale gas plays: The case of Texas. *Energy Policy*, 39(5), 2974–2981.

Texas State Historical Association. (2017). Carrizo Springs, Texas. Retrieved from https://tshaonline.org/handbook/online/articles/hfc02.

Texas Drilling.com. (2017). Summary of data near Carrizo Springs, Texas. Retrieved from www.texas-drilling.com/dimmit-county/carrizo-springs. Accessed on 6/9/2017.

Tunstall, T. (2014). Economic impact of Eagle Ford Shale, *Petroleum Accounting and Financial Management Journal*, 33(2), 11–22.

Wintergarden Groundwater Conservation District. (2007). Official district rules and regulations. Retrieved from www.twdb.texas.gov/groundwater/docs/GCD/wgcd/Wintergarden_GCD_Rules.pdf.

Yates, B. (2014, March 2). Fracking has made Carrizo Springs a boom town. *Liberty Voice*. Retrieved from http://guardianlv.com/2014/03/fracking-has-made-carrizo-springs-a-boom-town/.

7 Oil and Gas Industry's Responses to Communities Oppositions

Introduction

The communities that opposed fracking within or outside their city limits and expressed concerns over its negative externalities also met with representatives from the oil and gas industry from time to time. To collect information on specific measures that were undertaken by individual oil and gas companies to address communities' concerns proved to be quite a challenging task in the study. I had to contact various non-profit organizations in the state that represent the oil and gas industry and play an important role in addressing communities' concerns. Individuals from these organizations volunteered general information on the measures that are used to suppress communities' opposition to fracking.

In exploration and production of oil and natural gas in rural and urban areas, the oil and gas companies do not hesitate to send their own staff to talk to community residents or avail the services of non-profit organizations that provide support services to the industry including community outreach. Usually, the larger oil and gas companies have their own public affairs division with dedicated staff members whose responsibilities include the overseeing of community relations. They conduct meetings with community members to reduce or eliminate the community's opposition to fracking and establish a meaningful relationship with them to reduce obstacles in the production process that might cost the company money, while the non-profit organizations try to garner support for the industry using various tactics that include the involvement of community members in various outreach activities and the use of advanced communication technology. This chapter provides a glimpse into the strategies used by oil and gas company representatives to address community residents' concerns.

Oil and Gas Companies Efforts to Build Community Relations

There has been more opposition to fracking than traditional oil and gas drilling in the state. The varying degrees of ongoing opposition to fracking in a state where people are no strangers to drilling and the presence of rigs dotting the landscape conveys several important messages. First, fracking's environmental impacts are far more serious than those of traditional drilling practice, which has not evoked similar responses from community members. Second, there exists dissatisfaction among community residents with local ordinances and state regulations that are deemed inadequate in providing protection to people from the negative externalities of fracking. Third, affected individuals would continue to oppose fracking irrespective of the economic gains it generates until and unless their social needs receive due attention from the local and state government. Fourth, oil and gas companies need to look beyond building community relationships and internalize the social costs of production.

To address the aforementioned concerns, oil and gas companies in the state have acted both individually and collectively. In order to gain community support, they have made social investments in communities by sponsoring various festive events contributing to building public parks, enhancement of a community golf course, and other causes. These investments evoked mixed responses in communities – some helped to score support while others had little or no effect. In the latter type, individuals' environmental concerns overshadowed the gains from such investments in the community. Concerned residents' deep sense of losses in air quality, public health, and quality of life provided them with the impetus to oppose fracking within or near the city limits.

Communities' opposition to fracking has not stopped even with the decline in fracking activities in some parts of the state. In the Barnett Shale, anti-fracking sentiment still prevails. Community residents are fearful of fracking making a comeback in their communities. In these communities, strong leadership coupled with support from environmental organizations have helped local opposition groups to survive and keep their agenda alive.

Non-Profit Organizations' Efforts to Build Community Relations

To render support to the smaller oil and gas companies in building community relations, non-profit organizations have developed in various parts of the state. They offer to build a platform for public support and educate community residents on drilling facts and the safeguards that are in place to reduce or minimize damages to the environment. From the interviews conducted with representatives of these organizations, it has become clear that their scale of operation and mode of communication varied from one place to another. For example, the Barnett Shale Education Council (BSEC) offered community outreach services only in the Barnett Shale area of North Texas while Texans for Natural Gas (TNG) reached out to community residents in north, south, and other parts of the state including the Gulf coast.

On the other hand, the TXOGA is a large organization that looks after the interests of the oil and gas industry in the state. Established in 1919, this non-profit organization has members from all oil and gas companies doing business in the state. It partners with national agencies like the American Petroleum Institute, various states' oil and gas associations, and many others in building a support base for the oil and gas industry in the state.

Keeping the industry's interests in sight, TXOGA oversees the state regulations, legislations, judicial affairs, and public/industry affairs. Its public affairs division is responsible for the dissemination of statistics and other relevant information to the media for distribution to the public. Periodically, the organization sends out newsletters to its online subscribers and encourages state residents to become members to render support to the industry.

The non-profit organizations engaged in building community relations use two distinct approaches – traditional and non-traditional. The BSEC founded in 2007 uses a traditional approach. It played an important role when fracking reached its peak in the Barnett Shale area in 2012. It conducted face-to-face meetings with community residents at town halls, homeowners' associations' gatherings, Rotary Clubs, and various community events. In those meetings, the staff answered community residents' questions on oil and gas production using the fracking technique and efforts were made to allay their fears that may have stemmed from reading articles on the internet.

In educating community residents on the facts of drilling, the BSEC often used slide presentations followed by a question and answer session. In some communities, rig tours were organized with a detailed commentary on the drilling process. These tours were deemed quite effective in building community relations. At the drilling sites, BSEC staff showed individuals the safeguards that were in place to prevent leakages and reduce environmental effects. Proving the adage that seeing is believing, the strategy helped to ease many individuals' fears and change their mind.

When fracking first started in the Barnett Shale, BSEC conducted polls to collect information on communities' concerns. The information was then relayed to the oil and gas companies and they responded to them. They also made changes at the drilling sites in some communities in response to changes in the local ordinances. For example, the oil and gas companies installed different lights at the drilling sites to reduce glare, while sound walls were built to contain noise from the fracking sites.

The BSCE, as part of its outreach efforts and to keep the community apprised of the changes made at the drilling sites, published articles in the local newspapers and distributed flyers in the community to dispel people's misconceptions about oil and gas drilling. Such efforts were most noticeable in Denton, prior to the city's adoption of the ban on fracking.

On the other hand, TNG used a non-traditional approach in its efforts to garner support for local drilling. Being a relatively new non-profit organization that was established in 2014, it heavily relies on the use of social media to communicate with those individuals who support drilling in their community. TNG's choice of social media as an inexpensive mode of communication has helped to keep its operation costs low, made it possible to decentralize, and extended its area of service.

With the help of social media, TNG has built an online platform to provide support to those community members who often feel isolated or threatened by the presence of a strong anti-fracking group in their community. TNG's website helps to lend a voice to those supporters of oil and gas drilling in a community who are afraid to speak out at public meetings. Further, in its efforts to galvanize support for fracking in the community, TNG sends out regular e-mails to its members, tries to recruit new members online, and runs paid advertisements from oil and gas companies on its website that are relevant to its mission. To gauge public sentiments toward a new oil and gas drilling project in a community, it also conducts online signature petition drives to support such activities.

Conclusion

The communities' opposition to fracking has sometimes been interpreted by the oil and gas companies and non-profit outreach organizations as reflective of people's anti-fossil attitude or a propensity to create problems in the energy production process. At other times, such opposition has been interpreted as individuals' lack of information or adequate knowledge on the fracking process itself, which apparently is considered safe by the oil and gas industry. With reference to the first assumption, it is hard to believe that in a state where all community residents are highly dependent on the availability of cheap fossil fuel for their mobility needs, they would develop such an anti-fossil attitude. As for community residents' lack of information on fracking, that assumption lacks veracity. In an era of advanced technology where access to information is quick and with the media constantly feeding people with the latest news on fracking in various news outlets, it is hard to believe that people in an opposition group have not done their fact findings and checking while continuing to press the government at various levels with their demand for greater environmental and personal protection from the negative externalities of fracking.

Whatever maybe the opinion of the oil and gas companies, it has become evident from their dealings with those communities that oppose fracking that they prefer to use a piece-meal approach rather than a holistic one. Although the adoption of such a stance has been abetted by the weak and fragmented regulatory structure of the state and helps to save money, it does not allow the balancing of communities' welfare concerns with those of the oil and gas industry. Further the state's long history of oil and drilling has only proven to be advantageous for the oil and gas companies. It has helped to develop states' bias toward the oil and gas industry and does not require the oil and gas companies to internalize the social costs in the production of oil and natural gas. Under such circumstances, it comes as no surprise why communities continue to oppose fracking. To quell opposition, the welfare concerns of community members need to be balanced with those of oil and gas companies and this calls for meaningful changes not only in local ordinances, but also in the fragmented state and federal regulatory policies.

8   In Retrospect
Implications of Fracking Regulations and Communities Oppositions

Introduction

In Texas, oil and gas explorations started at the beginning of the twentieth century. During that time period, in the absence of well-defined federal regulations and amidst greater expectations of the need for the state to regulate, the state had gradually developed its own regulatory framework. With the expansion of oil- and gas-related activities and growing complexities in it, a myriad of issues became evident that had to be addressed in a fair and equitable manner. These issues ranged from land acquisitions to leasing and royalty payments, drilling of wells for oil, gas, and groundwater, and management of hazardous waste to disputes between the oil and gas companies and local surface and mineral rights owners. The states did realize that if they remained unresolved or unaddressed, more damages to both private and public interests were likely to occur that might pose obstacles in the harnessing of energy from fossil fuels. Also, it would be likely to impact on the profitability of the industry and affect both the state economy and that of some local governments, where it is harder to attract investments and create new jobs.

Reflections on State Regulations on Fracking

Over the years, Texas has gradually developed its own regulatory framework to control oil and gas interests. It bears most of the responsibilities to control oil and gas drilling including fracking. A comparison of the state's regulations on fracking with those of other states like Oklahoma and North Dakota has shown that they are more detailed and mature than those of the latter two. That is not surprising because in those states where oil and gas production are relatively recent activities, the regulations are still in a nascent stage of development. Despite Texas' edge over other states in regulatory matters over fracking, questions still remain over their efficacy in the balancing of private interests with those of the public. The favored stance of the state toward the oil and gas industry has created discontent among many people and led to opposition to fracking in some communities.

A review of the land laws reveals a bias toward mineral ownership and preference for the development of mineral wealth in the state. The state, in its commitment to develop mineral wealth and reap economic gains from it, has removed some of the hurdles in drilling on private lands. For example, it has bestowed a predominant status on mineral rights owners while in its split estate system, the surface landowners have not always been treated fairly. Often the surface land owners with limited rights have had to surrender their land to drilling if the mineral rights owner has entered into a lease agreement with an operator. Also, in cases of dispute over the taking of lands for drilling purposes, courts have mostly ruled in favor of mineral rights owners and the energy industry. Though compensations have been paid by the oil and gas companies to the surface land owners for their losses arising from the use of land for drilling, the monetary amounts are often deemed inadequate in offsetting corresponding losses.

In other cases, where plaintiffs living close to fracking sites have shown evidence of damages to life and property due to drilling, the courts' decisions have been unpredictable. There exists arbitrariness in courts' decisions in estimation and evaluation of the losses of individuals from drilling activities, resulting in similar cases being awarded dissimilar amounts. Further by reducing the awards amount on subsequent and similar cases, the courts have sent a message to those plaintiffs who have won their cases and individuals living in communities close to fracking sites that energy development takes precedence over all other interests in the state. This attitude has been partly responsible for the overlooking of damages incurred on properties and surrounding areas and not treating them as a consequence of the negligence of the operators, rather as a part and parcel of the drilling process that therefore needs to be tolerated for its economic benefits.

Even the state's Rule of Capture that permits the capture of oil and gas by operators from adjacent private property owners without their consent or compensation is another law biased toward energy development. It seems to carry a punitive burden on those who do not consent to fracking in their private property by inflicting double losses to those landowners whose properties lie atop an oil patch. These double losses arise from the taking of a fair share of mineral wealth without compensation and in a declining quality of life from fracking in the adjacent property. In a business environment, the capture rule may sound justified as it aims at promoting efficiency in the oil and gas production process by not letting energy resources go to waste. Unfortunately, such seizure of mineral wealth without owners' consents amounts to state-allowed violations of mineral ownership rights of those property owners who are reluctant to sign a lease with oil and gas companies for fracking purposes. By establishing the Rule of Capture, the state has not only allowed the relaxation of regulations on the oil and gas industry but also has assumed the role of a serious stakeholder in the industry's profitability. The RRC, the primary regulatory authority over the state's oil and gas industry, has not offered a remedy yet to offset the capture process. This is perhaps because the state's Rule of Capture is in alignment with its mission statement – to enhance development and promote economic vitality in the state.

When it comes to the use of water in the fracking of shale rocks, the state's regulations on fracking seem inadequate too. In most cases, wells drilled to withdraw underground water for fracking do not require a permit, and there exists no limit on the withdrawal amount. Only surface water use in fracking is subject to regulations by the TCEQ. Since fracking also takes place in many arid areas of the state where water needs to be hauled from distant places, this means that less water is available for agriculture and other competing uses in the origin place of the groundwater supply. To deflect public attention and concerns over the large volume of water used in fracking in areas of high-water stress, the oil and gas industry has claimed that its total water use at the state level is minimal in comparison with other water intensive activities like agriculture and ranching. But the ramifications are large at the local level. This warrants a change in the state's water code and the inclusion of fracking as a water intensive production activity. Such an amendment will help to end controversies and enable the water districts to develop their own guidelines on withdrawal of water in their water management plan. This would help to address the inequities in water distribution in arid regions at a time of water stress, without hurting the interests of one production activity over another. Another aspect that needs equal attention is the recycling of water, where possible to reduce water use. Although recycling is done at some fracking sites in the state, the overall figure is quite low in comparison with other states like Pennsylvania. This calls for the strict enforcement of recycling by all big and small oil and gas companies and making the infrastructure available to make this possible and not leaving recycling as an option for interested companies to explore.

Closely associated with fracking is the issue of environmental injustice. Fracking has disproportionately inflicted a higher level of harm on those communities that are located closer to the fracking sites and often populated by minorities with lower incomes, as observed in the Eagle Ford Shale. Here, the victims of fracking do not know where to lodge their complaints or how to rally for more protection from fracking. The local governments in many of the smaller rural communities are equally ill prepared to deal with such issues of equity concerns (Johnston, Werder, & Sebastian, 2016). They not only lack the capacity to address it but are also afraid to take any actions that might drive away those oil and gas companies that provide the much-needed jobs in the local economy.

There exist multiple factors that influence the state's tepid control on oil and gas drilling activities. First, the role played by the oil and gas industry and its lobbying partners cannot be ignored. They constantly remind the state's elected officials of the industry's contribution to the state and local economies in terms of employment and revenue. If stringent regulations are adopted, oil and gas companies' profits would decline with the increase in costs of operation. Second, the state's elected officials, including the commissioners of the Railroad Commission (RRC), receive monetary contributions from the oil and gas industry during their election campaigns. These contributions interfere with their abilities to regulate properly (Hasmeyer, Weider, & Suderman, 2014). Third, an open-door policy allows the recruitment of people from the oil and gas industry to serve in various capacities in the RRC. This can result in some degrees of hesitation to regulate the industry that once provided for their livelihood. Fourth, both the RRC and TCEQ have in common a single mission statement that focuses on the economic vitality of the state. This stance conflicts with their regulatory attitude and behavior. Furthermore, it makes the TCEQ the only environmental regulatory agency in the nation to even have such an economic mission statement while advocating environmental protection.

Finally, the role of the state's culture cannot be overlooked in undermining the will to enforce regulations on the fracking related activities of the oil and gas companies. The state's political culture has helped to inculcate among the state's regulatory agencies the tendency to prioritize business interests over those of the environment. Its evidence manifests in the forms of opposition to the EPA, granting of state exemptions to oil and gas companies, and passage of a state bill that favors the industry. For years, the state has resisted the EPA's tougher standards on emissions of methane and other gases blamed for air pollution and climate change. It has not hesitated to get involved in costly legal battles with the EPA, protesting against the agency's requirements of tougher standards on industries that will increase their costs of production and lower profit margins in the long run while trying to provide more protection to the people.

The impact of culture is also evident in the state's leniency on adjustments in setback requirements. The RRC has made available several exemptions to oil and gas companies. One such exemption allows the oil and gas companies to drill wells very close to adjacent property if they can prove that a selected site is the most efficient one. In granting such an exemption, what has been overlooked is the high density of gas emissions detected and recorded at fracking sites that pose health risks to those individuals living closer to fracking sites (Meng & Ashby, 2014). With reference to the disclosure of names of chemicals used in the fracking fluid, the state has provided protection to companies under the trade secret clause. It enables companies to withhold the names of certain chemicals that give them an edge over another. Emergency responders may know the names of some of these chemicals to strategize their action plan but the lack of community awareness make individuals at close proximity to fracking sites more vulnerable to exposure to toxic chemicals in the event of accidents. Although companies have assured that safety measures are undertaken in dealing with such chemicals, accidental spills or their sudden release into air, soil, and waterways cannot be ruled out. This tendency to withhold information from the public not only fails to satisfy their Right to Know but also undermines their trust in the regulation process (Konschnik, 2014).

Another impact of culture can be observed in the state's willingness to protect the oil and gas industry's interests through the passage of the House Bill 1794. Passed in 2016, the bill has made it tougher for local governments to sue an oil and gas company for polluting the air or water and caps payout limits of industries at $2 million. This also indicates the capture of the regulatory agency in the state by the oil and gas industry. Intense lobbying by the industry along with the presence of a state political culture that is highly conducive for oil and gas exploration and production are partly to be blamed.

It is under such circumstances of unabated state support to the oil and gas industry that individuals' requests for adequate protection from the externalities of fracking have gone either unheeded or received scant attention. The problem has been further accentuated with the revelation of RRC's failure to inspect wells in a timely manner. Funding shortages have hindered the agency in the hiring of inspectors. In the absence of timely inspections, some wells have not been inspected for five or more years, endangering those individuals who live near the fracking sites (Handy, 2017).

Reflections on Communities Oppositions to Fracking

In the oil and gas producing state of Texas, people are quite used to living in close vicinity to oil and gas production sites and seeing oil rigs towering over the trees in the landscape. With the advent of fracking in the state, not all people have felt the same about this type of shale drilling. Fracking has evoked far more complaints from people than from those who live close to production sites using the traditional method of drilling. The individuals from communities located closer to fracking sites have condemned it for various reasons despite their knowledge of the many economic benefits this type of drilling has rendered to the state and local economies.

Based on individuals' own experiences with fracking, opposition from communities has varied in magnitude. Some communities adopted a strong stance against it while others have displayed only a moderate to mild opposition to it. The differences in the level of opposition can be partly attributed to the locus of fracking activities, which in turn influenced the severity of the environmental problems emanating from this drilling activity. The communities that experienced fracking within the city limits, as observed in the city of Denton, displayed a higher degree of opposition due to several environmental risks. This contrasts with those communities where fracking operations took place outside the city limits, as observed in the cities of Irving and Carizzo Springs, who displayed a moderate to mild level of opposition. The severity of environmental impacts also influenced the types of actions taken by individuals to restrict this activity when local ordinances and state regulations were considered inadequate. Whatever may be the level of opposition to fracking in a community, individuals' decisions to oppose this activity were based on their perceptions of risks and evaluation of its effects from a reference point of quality of life.

A closer look at communities impacted by fracking has revealed that it has created a certain amount of divisiveness in them. Support for this activity came from those who benefited from the royalty payments and well-paid jobs in the oil and gas companies and they also ignored fracking's environmental impacts. Others overwhelmed by its negative externalities opposed it. They complained about fracking and sought more protection from the local government. In a Republican majority state, where most people favor fracking as shown by various poll results, this type of behavior has not come as a surprise. In Denton County, where approximately 66 and 69 percent of registered voters had voted for Republican candidates in the 2008 and 2012 general elections, the majority of residents in the city of Denton did not hesitate to oppose fracking in their community. They went as far as passing a ban on fracking but this was only short lived. The declaration of the fracking ban shocked and surprised many people both within and outside the state. It did send a message about the seriousness of fracking's environmental problems that many people found intolerable even after residing in a state where oil and gas producing activities are rampant.

Impacts of Environmentalism

From the individuals' opposition to fracking it has also become apparent that there exists the prevalence of a strong spirit of environmentalism that has overshadowed nimbyism in communities (Christopherson & Rightor, 2014). Though both the concepts share similar values toward environmental stewardship, there exists a subtle difference between them. Environmentalism refers to individuals' protective sentiment toward the environment while nimbyism refers to the "not in my backyard" syndrome. The difference lies in their scale of opposition. In nimbyism, usually a smaller group of people comprising of those who live either very close to a source of pollution or are likely to be impacted from a planned project that will pollute, oppose the actual or likely environmental damages. In environmentalism, however, the opposition comes from a larger group of people. It comprises of not only those who have been adversely impacted by the negative externalities of any production process but also others who detest activities that cause environmental deterioration (Smith, Michaud, & Carlisle, 2004).

The groups that have developed in the communities to oppose fracking are akin to grass-root organizations. Individuals in these groups have dared to adopt a stance against the big and small oil and gas companies that are located both in and out of the state. From their opposition, hard truths on environmental pollution from fracking have emerged, like the earthquakes, damages to transportation infrastructure, increase in crashes and fatalities as a result of the increase in heavy trucks on roads, and water shortages in arid areas. Although the opposition groups have not always been successful in achieving their desired goals of restricting the effects of fracking in their community and seeking more protection through local ordinances and state regulations, this has not discouraged them from rallying for their cause or caused them to cease to exist. Many of these community opposition groups have joined the Texas Grassroots Network to form a larger coalition group and maintain links with other communities in the nation with similar problems and objectives.

Further, the opposition groups have continued to participate in activities to exert pressure on their local representatives for more protection. For example, to regain local control over fracking, the communities' opposition groups have provided support to state politicians like Terry Canales, a state representative from the San Antonio area. He filed House Bill 3403 in the state's legislature in March 2017 to provide protection to school children from fracking. The proposed bill will not allow the drilling of new oil and gas wells within a radius of 1,500 feet from the property line of schools and child care facilities both public and private and with at least 100 children. If approved after tough opposition, this bill would require many cities like Fort Worth to revise their current ordinance that requires a much smaller setback for oil and gas drilling. A similar bill has also been filed by state senator Judith Zaffirini in response to residents' pleas for more protection from fracking in the Eagle Ford Shale. This bill also requires a 1,500-foot setback for drilling purposes, but it requires a hearing before the RRC prior to making any decision (Baker, 2017).

Outcomes of Community Opposition

It is evident from the case studies that oppositions to fracking from communities do not match in scale and synergy. Nevertheless, collectively they have helped to attract the attention of the state's regulatory officials and produce both favorable and unfavorable outcomes. The favorable ones include the passage of local ordinances to control noise and use of bright lights at fracking sites and arrangements with oil and gas companies to develop a separate route for the movement of heavy truck traffic to avoid residential neighborhoods. Individuals' complaints have made the local governments enforce zoning ordinances more to prevent the haphazard siting of drilling wells and make revisions in the setback requirements for the drilling of new wells. The oil and gas operators have resisted the changes in local ordinances and when unsuccessful have sought the help of a grandfather clause to make the local government honor the old contracts.

Other favorable outcomes include the state's approval of funding for the installation of expensive air quality and seismic monitors. If Denton and Irving residents and those from the surrounding communities had not complained to the RRC officials, little or no action would have been taken by this agency to monitor air quality or study the swarms of earthquakes that hit this region. With reference to the latter, the agency would have held steadfast to its belief of no possible connection between injection wells and sudden earthquakes, in the absence of credible scientific evidence. Under public pressure, the state has approved the funding of scientific investigation to determine the cause of earthquakes near injection wells. It also made amendments in requirements for the citing of new disposal wells and controlled injections in those disposal wells which stand at a higher risk of inducing seismic activities.

In Fort Worth, the residents' numerous complaints about the unsafe driving conditions on the roads and highways due to increases in truck traffic and other traffic related problems have drawn attention to various transportation related matters in energy development. To address them, administrative solutions were required. The problems called for the local governments to involve stakeholders from the community and the industry to develop a joint plan for road safety and to set aside special funds collected from oil and gas companies for the costly road repairs. In some cities, though such a type of fund is available, the small amount is inadequate to undertake costly repairs. In another case, water shortages in the city of Carrizo Springs that affected agriculture and ranching have drawn attention to the need for the development of a more balanced water management plan. Since the ambiguities in the state's water code hinder the equitable distribution of groundwater, they must be revisited to remedy the situation.

The unfavorable outcomes include various types of externalities ranging from air pollution to the damage of roadways from fracking related activities that have taken a toll on individuals' quality of life as observed in the Eagle Ford Shale area. Here, rural residents' complaints have gone mostly unheeded, while opposition from an urban community in the Barnett Shale area has led to the passage of the House Bill 40. Triggered by Denton's fracking ban and passed at the behest of the oil and gas industry, the bill delivered a big blow to the authority of local governments in the state. Prior to Denton's fracking ban, several other cities in the state had passed a ban on drilling. The state tolerated the previous bans because there were no drilling activities within those cities to harm the interests of the oil and gas industry. Moreover, these bans were passed to prevent future drilling activities within the city limits. The HB40 made all such drilling bans passed by local government unenforceable.

The bill serves as a constant reminder to all "home ruled" cities in the state to be cognizant of their regulatory authority. As per the state's home rule charter, a home rule city can act in the best interest of the local population, but it cannot adopt a regulatory stance on an issue if the state government has not shared its regulatory power with local governments. Under such circumstances, a city-imposed ban has no legal standing. The HB40 also conveyed to all local governments that any threats to a state authorized activity, like drilling and the production of oil and gas irrespective of the methods used, will not be tolerated. The passing of the HB40 serves as another indication of the state's preferential treatment of the oil and gas industry and that economic interests take precedence over other interests of the public and the environment.

Attitudes Toward Oil and Gas Production

In an era where renewable forms of energy are gaining grounds with increases in their market share of energy production, individuals' attitude toward the oil and gas industry has changed over the years. People do not look at the oil and gas industry the same way they once did when it was the sole source of energy. The strong media has created greater awareness of its numerous oil spills in the coastal waters and on land that have caused a multitude of damages to the environment and individuals' lives and properties. Adding to the industry's negative perceptions are the documentaries like *Deepwater Horizon* and news on Exxon Mobil's efforts to hide information on climate change along with that of the layoff of thousands of people during the oil bust of 2016, as well as the lack of efforts by most oil and gas companies to solve environmental problems in order to make people's lives better. Taken together, they all have contributed toward the low social approval of the oil and gas industry.

The people's loss of trust in the activities of the oil and gas industry have been reflected in the results of a survey that was conducted by a consulting firm called EY. According to EY's 2017 surveys of average people and oil and gas executives, even though 75 percent of the industry's executives rated themselves as good corporate citizens, only 37 percent of the people thought that the industry can be trusted to do the right thing. Also, with people viewing renewable energy far more favorably than oil and gas, only 14 percent of surveyed respondents had shown support for oil and gas as the sole source of energy, while 29 percent did not, and the rest were willing to tolerate oil and gas until alternatives are found (DePillis, 2017).

Amidst the ongoing conflict between communities' opposition groups and the oil and gas companies, The Academy of Medicine, Engineering and Science of Texas (TAMEST) released its Shale Task Force Report in 2017. The report titled, "Environmental and Community Impacts of Shale Development in Texas," has been produced by and with contributions from experts in academia, environmental organizations, the oil and gas industry, and state agencies. It has attested not only to the state's and individuals' economic gains from the fracking of shale resources in the state but also to the threats that this unconventional drilling technology poses to land, air, surface water, transportation infrastructure, and other natural resources.

If fracking is to be continued in both urban and rural communities, the oil and gas industry needs to make fracking more safe and secure. In the twenty-first century, business ethics call for social responsibility among producers of goods and services. Keeping this in mind, those oil and gas companies producing energy from shale resources need to pay more attention to climate change, social costs of production, and make efforts to integrate them into the total cost of the production process.

The Future of Regulations on Fracking

In an era of deregulation under the Trump administration, states have been asked to refrain from the implementation of those regulations that will reduce industries' profitability and lead to job losses in the local and state economy. In such a national call for deregulation, Texas like many other oil and gas producing states in the nation, now has less incentive to regulate oil and gas drilling activities such as fracking. The current political climate has signaled the opening of a new policy window to bring about changes in regulations. This time, the policy changes will involve the relaxation of those existing regulations that are oppressive for industrial growth. The larger oil and gas companies in the state have shown an indifference toward deregulation as they have already invested in costly pollution abatement measures. The likely beneficiaries would be the smaller oil and gas companies, who have welcomed deregulation. For example, under the Trump administration the EPA has announced that it will no longer need companies to collect information on methane emissions from their oil and gas production sites. The relaxation of such a law passed by the previous administration to limit methane emissions has been lauded by the smaller oil and gas companies. They do not need to invest in new environmental friendly technology and can expect to see an increase in their profits without worrying about environmental pollution.

Along with deregulation has come the federal government's announcement of efforts to reopen the coal and nuclear power plants in the nation that could no longer operate because of higher costs of operation and emissions of pollutants. Their re-openings are expected to create jobs and ensure both reliability and stability in power production and supply. These initiatives can be deemed as mixed blessings for the oil and gas industry. Though deregulation is likely to positively impact the industry, the reopening of the coal and nuclear power plants will lead to a decline in the demand for natural gas from the utility industry. Since fracking started, the oil and gas companies have been selling natural gas to the utility plants. The cheap and abundant supply of natural gas has helped many coal-based utility plants to substitute coal with this fuel and cut back on harmful emissions of carbon dioxide blamed for climate change.

From 2012 to 2016, fracking has been partly responsible for the closure of more than 500 coal-based utility plants (Osborne, 2017). Their reopening would mean a significant decline in natural gas' market share as a clean fuel for electricity generation along with a shift in emphasis from renewable energy like solar and wind to coal and nuclear. Even the public may not like the change. Used to monetary savings from the production of low-cost energy generated by natural gas and renewables like solar and wind, only a few may be willing to give up such savings and tolerate the hike in energy costs for the sake of small job gains in the economy.

In addition, the EPA's announcement of a plan in October 2017 to repeal the Clean Power Plants that was introduced during the Obama era was well received by the nation's coal-fired power plants. In Texas, where the existing coal power plants supplied only 30 percent of the state's electricity needs during June 2017, this seemed like an opportune moment for the state's coal power plants to make a comeback and regain their market share. Unfortunately, this is not possible since most of the existing plants are struggling to survive in a competing environment that has been made worse with contributions to the grid from lower priced wind and solar energy (Handy, 2018) along with that produced from natural gas. Under such changing economic conditions, the state's largest energy producer, Vistra Energy, announced its plans for the closure of one of its coal power plants in the state (Friedman & Plumer, 2017).

Although such an announcement offers a glimmer of hope in the possible reduction of harmful emissions from power plants, it has failed to satisfy environmentalists and community opposition groups. With some of the current state regulations on fracking already lacking enforcement, they fear that any further relaxation in a volatile regulatory climate would only add to the social costs of production of oil and natural gas in the state. And this is a serious concern since these costs are disproportionately borne by those individuals who live closer to the fracking sites.

With deregulation and the reuse of coal and nuclear fuel for electricity generation, there lies a greater threat to individuals' welfare from environmental pollution. This has posed additional concerns among many individuals who are already susceptible to the negative externalities of production, including fracking. It is during such times that communities' opposition groups along with the watchdog environmental organizations can assume additional responsibilities of protecting the environment.

The larger environmental watchdog groups, fearing losses in environmental quality, often try to hinder federal agencies' actions like the EPA's attempts at deregulation, by filing lawsuits. In the legal battle, the endangerment clause is often employed to disprove evidence that led to the adoption of the regulation in the first place along with the proof that under present circumstances, such protection is no longer necessary to ensure public safety and welfare. Since the process takes a long time, it helps to delay deregulation. The fight to maintain an existing regulation that may have been considered inadequate in the past often makes people realize its value in the current context, especially when individuals' welfare is at stake.

Conclusion

No matter what the political, economic, and social climate may be, the balancing of private interests with public ones remains essential all the time in fracking, which will continue for years to come with support from the federal and state government. Undoubtedly this task is a challenging one but attempts can be made with the help of partnerships and collaborations. For example, the partnership called the Shale Task Force Committee of The Academy of Medicine, Science and Engineering of Texas was developed to scrutinize the contentious issue of fracking in the state and has helped members reach a consensus and acknowledgment of those environmental problems that were unforeseen at the beginning of fracking along with the need to address them. The committee has made public its findings in the form of a public report that can be downloaded from its website.

The scope of such a type of partnership needs to be expanded beyond the publication of a fact-finding report. By including community members and local government officials representing the diverse interests of the community and who can also provide an insight into the problems arising from fracking, a stage can be set for a meaningful dialog between state officials, oil and gas companies, community residents, and independent researchers. In such a dialog, the pros and cons of various policy options that can help to address the problems can be discussed and debated. Once a consensus is reached, there will emerge those policy recommendations that are reasonable and would face least resistance in adoption by the state and local government.

Another option that needs to be explored is the state's sharing of its regulatory authority over fracking with local governments. Under the current preemptive stance adopted by the state government, discontent among community members is not likely to subside but continue and may even escalate if fracking gains momentum in the future with the rise in oil and gas prices following an uncertainty in the supply of energy in a volatile political climate and both natural and manmade disasters. A shared governance approach can help local governments make judicious decisions since they know best the problems in their jurisdictions. By using such an approach that calls for the involvement of community members, perhaps community opposition to fracking can be reduced and a balance can be struck between private and public interests.

Business ethics require all companies to pay attention to social responsibilities; it is therefore important for oil and gas companies to address the distinct social needs of communities where they operate and internalize the social costs of production. For instance, they need to pay for their share of damages to those property owners who have been affected by fracking induced seismic activities instead of denying them and accusing those people who oppose fracking in the community of being anti-fossil fuel activists. Failure to do so and by trying to escape from their social obligations, the oil and gas companies will not be able to cultivate a good social image or reduce opposition in communities. A behavioral reform with greater acceptance of social responsibilities along with investment in efforts to reduce the externalities of fracking hold the promise of building better relations with communities and a reduction in individuals' opposition to fracking.

Since all companies cannot be fully trusted to undertake social actions on their own, the state government needs to make good use of both incentives and regulatory mechanisms to address the social concerns of communities near fracking sites. Through incentives, it can encourage the oil and gas companies to invest in environmental friendly technology and fund innovations while rules and regulations can help to reduce toxic emissions from fracking sites. In achieving the latter objective, the state needs to rethink its bias toward the oil and gas industry. The state legislature, with a predilection toward oil and gas production, rejects the rules that are strongly opposed by the oil and gas industry and this obstructs the adoption of rules by environmental regulators in response to public concerns.

Since the state's pro oil and gas attitude continues to interfere with the balancing of public interests with those of private ones and enables the capture of the state's regulatory agency, the public and environmental organizations need to take a more proactive stance against the externalities of fracking in the state. Together they can exert more pressure on the state government to bring about meaningful regulatory changes on fracking and advocate the return of power to local governments, as they are most suited to addressing the needs of the local population.

References

Baker, M. (2017, March 17). Bill seeks to reclaim local control over oil and gas drilling near schools. *Star Telegram*. Retrieved from www.star-telegram.com/news/business/article139106428.html.

Christopherson, S. & Rightor, N. (2014). Nimbys or Concerned Citizens? *Progressive Planning, 108*, 32–35.

Denton County. (2017). Election results. Retrieved from www.votedenton.com/election-results/#PastElections.

DePillis, L. (2017, May 14). Oil and Gas industry knows it has an image problem. *Houston Chronicle*, B3.

Friedman, K. & Plumer, B. (2017, October 10). Repeal of climate plan sets up a bitter battle. *Houston Chronicle*, A1 & A11.

Handy, R. (2018). Added Texas solar capacity may burn power companies. *Houston Chronicle*, March 18. 2018, B3.

Handy, R. (2017, February 9). Thousands of oil rigs go unmonitored. *Houston Chronicle*, A1 and A12.

Hasmeyer, D., Wieder, B., & Suderman, A. (2014). Texas officials turn blind eye to fracking industry's toxic air emissions. *Inside Climate News*. Retrieved from https://insideclimatenews.org/news/20140218/texas-officials-turn-blind-eye-fracking-industrys-toxic-air-emissions.

Johnston, J. E., Werder, E., & Sebastian, D. (2016). Wastewater disposal wells, fracking, and environmental injustice in southern Texas. *American Journal of Public Health*, 106(3), 550–556.

Konschnik, K. (2014). Goal-Oriented Disclosure Design for Shale Oil and Gas Development. *Natural Resources Journal*, 54(2), 319–359.

Meng, Q. & Ashby, S. (2014). Distance: A critical aspect for environmental impact assessment of hydraulic fracking. *The Extractive Industries and Society*, 1(2), 124–126.

Osborne, J. (2017, September 30). Perry seeks to protect coal, nuclear plants. *Houston Chronicle*, B1 & B5.

Smith, E., Michaud, K., & Carlisle, J. (2004). Public opinion about energy development: Nimbyism vs. environmentalism. Prepared for delivery at the *Annual Meeting of the American Association of Public Opinion Research, Phoenix, Arizona*.

Index

Page numbers in **bold** denote figures, those in *italics* denote tables.